PRACTICAL, MADE EASY GUIDE TO ROBOTICS & AUTOMATION
[REVISED EDITION]

KERWIN MATHEW

PREFACE

Robots will become more and more intelligent and will take over more and more tasks that human beings find boring, dangerous or just humanly impossible. Unlike human beings, robots are devoid of feelings or emotions, not demanding and tireless. They will just perform the tasks as instructed or programmed without complaining.

Many factories and plants engage robots for such tasks as assembly work, welding and spray painting. To better understand and more fully appreciate the usefulness of robots, we have the need for a useful reference text. This book will help to fulfill this need.

Kerwin Mathew, Ph.D., PE, CMfgT, CPM

CONTENTS

1 INTRODUCTION

Industrial robots are reprogrammable multi-functional manipulators which are designed to move parts, tools or specialized devices for the performance of a variety of tasks. They are the new generation of "steel collar" workers capable of working three shifts a day without tea break and vacation and are immune to fumes, heat, radiation and other hazards.

With the introduction of microprocessor technology, industrial robots have become more and more intelligent and yet less expensive. As a result, they have become more acceptable to small and medium sized factories, taking over jobs such as materials handling, welding, spray painting and repetitive assembly work from human workers. They have revolutionized industries throughout the world. In their efforts to increase manufacturing productivity through increasing mechanization and advanced automation, industries will find the application of industrial robots a possible means for reducing costs, improving product quality and being less dependent on labor which has become scarce in supply.

Applications of such automated systems will require knowledge and understanding of how robots can be creatively utilized to solve the various problems concerned with manual operations and productivity.

2 WHAT ROBOTICS IS

A hundred years ago, no one had ever heard the word "robot". The term was first used by a Czechoslovakian writer, Karel Capek, in the 1920's. Capek wrote a play about a scientist who invents machines that he calls robots, a term which is derived from the Czech word "robota" meaning "slave-like work". He gave them this name because they were used to perform very boring work. At the end of the play, the robots kill their human owners and take control of the world.

Though there are now many robots in existence, these robots are quite different from the robots of science fiction films and books. Real robots are just machines controlled by a computer to work in a set way instead of being frightening, super-intelligent metal people. They are basically deaf, dumb, blind, have no sense of touch, smell or taste, have difficulty moving around, and have no intelligence of their own. However, with advances in microchip technology robots are now made with sensors, e.g., a TV camera "eye" or a microphone "ear", which provides them very limited senses like electronic sight and hearing.

Robots are used to perform many tasks, often tasks which are very dangerous or tiring for people to carry out such as welding car bodies. Being often able to work more efficiently than people, they are useful in factories. Robots never need holidays, sleep or meal breaks though they sometimes break down. As they can be re-programmed to do different jobs, some factories prefer to use robots rather than other automatic machines.

Some robots are used to perform tasks that would be impossible for people to carry out, such as working inside the radioactive section of a nuclear power station or exploring distant planets. Others, e.g., the small micro-robots used with a home computer, are just for fun or for learning about robotics.

Robots are capable of doing many different things, particularly in factories, where they are carefully maintained and organized to work alongside other automatic machines. As it is much more difficult to get them to work away from the ordered environment of a factory, robots are rarely used to perform outside, field work.

Though science fiction robots are often made to look human the appearance and ability of an industrial robot actually depend on the kind of task it has to perform. The majority of industrial robots are like "arms" bolted to the floor as the work they do can be carried out standing in one place. These arm robots are often referred to as manipulative robots as they hold or grasp things, e.g., a tool.

Arm robots are most commonly found in car factories though they can also be found in many other industries, e.g., clothing, confectionery, engineering and electronics. They are most suited for jobs that involve doing the same thing over and over again.

WHAT ROBOTS CAN DO

[1] [a] Feature Class:
Mobility
[b] Currently Available Capabilities:
Fast, efficient on smooth surfaces (wheeled systems) synchronized with all moving sub-systems (e.g., hand-arms, hand-eyes).
[c] Research Being Conducted:
Walking rough terrain propulsion, self-navigation systems, programmable omni-directional mobile systems.

[2] [a] Feature Class:
Manipulation
[b] Currently Available Capabilities:
Continuous-control articulations between base and gripper, point-to-point control, repeatable 1/8-inch positioning accuracy, ability to handle more than 350 pounds.
[c] Research Being Conducted:
General-purpose hand-arms and wrists and grippers, multiple hand-arm coordination, improved repeatable positioning accuracy, miniature and microscopic manipulators.

[3] [a] Feature Class:
Sensing
[b] Currently Available Capabilities:
Force and torque sensing, 2-D vision with discrete (mostly binary) recognition, primitive voice input.
[c] Research Being Conducted:
3-D vision with chrome (color, intensity, brightness) discrimination and depth and distance cueing, voice communication, positional sensing, and multiple-sensory input coordination and control.

[4] [a] Feature Class:
Learning
[b] Currently Available Capabilities:
Programmable (online and offline), teach/act modes, on-board and remote database access.

[c] Research Being Conducted:
General-purpose robot programming languages, deductive/inferential learning.

[5] [a] Feature Class:
Decision-Making
[b] Currently Available Capabilities:
Program-based, input-activated.
[c] Research Being Conducted:
Knowledge-based "expert system" model of working environment.

[6] [a] Feature Class:
Reliability
[b] Currently Available Capabilities:
For industrial applications, at least 500 hours mean time between failures.
[c] Research Being Conducted:
Self-diagnostic, fault-correcting.

3 FACTORS TO BE CONSIDERED FOR INTRODUCTION OF INDUSTRIAL ROBOTS

INTRODUCTION

Several techniques could be adopted for investment appraisal, each manager tending to adopt his own favorite method. It should be borne in mind that all appraisal results depend on the accuracy of the input data, some of which has to be estimated or forecasted. The most satisfactory situation arises when a relatively simple calculation shows an investment to be worthwhile by margins which exceed all possible doubts arising from shaky estimates. Fortunately, robotics often yields such positive predictions.

Whatever investment appraisal method is used, the input data for both savings and costs have to be accumulated.

CHECKLIST OF ECONOMIC FACTORS: COSTS AND BENEFITS

The headings which follow provide a framework for management analysis of the costs and benefits of the robotics installation:-

[1] Costs:
> [a] Purchase price of the robot.
> [b] Special tooling.
> [c] Installation.
> [d] Maintenance and periodic overhaul.
> [e] Operating power.
> [f] Finance.
> [g] Depreciation.

[2] Benefits:
> [a] Improvement of economic, technical and social influence.
> [b] Increase in output.
> [c] Improvement of quality.
> [d] Better utilization of existing plant equipment.
> [e] Greater flexibility in production.
> [f] Upgrading of technological innovations.

CHECKLIST OF COSTS

The following observations add perspective to the checklist headings.

[a] Purchase Price of the Robot

The purchase price of a robot varies greatly, especially if one's definition of robot includes simple pick and place devices with few articulations. The price range depends on the complexity and sophistication of the robot, e.g., weight handling capacity, control sophistication, sphere of influence and number of articulations. The higher priced robots are generally capable of more demanding jobs and their control sophistication enables them to adapt to new jobs when they have completed their original assignments. The more expensive and more sophisticated robot will also normally need less special tooling and incur lower installation cost. Some of the pick and place robots are merely adjustable components of automation systems, e.g., a popular model rarely contributes more than 20% of the total system cost.

[b] Special Tooling

The tooling may include transformers, clamps, an indexing conveyor, weld guns and a supervisory computer for a complex of robots which perform spot welding work on automobile bodies. The special parts presentation equipment may cost much more than the cost of the robot equipment, for assembly automation.

Unlike conventional machine tools, robots are not stand-alone equipment. Customers often request the robot manufacturers to bid on a turn-key basis because the interface with the workplace can be critical to success. For economic evaluation, the two main cost factors, namely, the cost of the robot and the cost of special tooling, are hence combined.

[c] Installation

Though installation cost is often regarded as an overhead because plant layout changes would occur anyway, it is sometimes charged fully to a robot project. A model change usually results in installation costs even if equipment were to be manually operated. There is no point in "penalizing" the robot installation for any more than a differential cost inherent in the process of robotizing.

[d] Maintenance and Periodic Overhaul

There is a need for regular maintenance - a periodic need for more thorough overhaul and a random need to rectify unscheduled downtime incidents - to keep a robot functioning in tip-top condition. As a rule of thumb a well-designed production equipment operating continuously for two shifts would incur a total annual cost of ten per cent of the acquisition cost.

This annual cost would vary of course, depending on the demands of the job and the environment, e.g., maintenance costs in a foundry would be greater than those in plastic molding.

[e] Operating Power

Operating power is computed easily through the product of average power drain times the hours worked. This is not a major cost of robotizing even with increased energy costs.

[f] Finance

In some cost justification formulas the current cost of money is taken into account. In others an expected return on investment is used to determine economic viability.

[g] Depreciation

Like other equipment robots have a useful life and it is a practice to depreciate the investment over the period of this useful life. As a robot tends to be general-purpose equipment, its useful life functioning multi-shifts should conservatively span eight to ten years.

Straight-line depreciation is most commonly used for costing purposes but the taxation schemes of different countries may involve depreciation weighted to early years.

CHECKLIST OF BENEFITS

Robots also bring a number of benefits which are not normally associated with other capital investments. In evaluating a proposed robotizing project, these benefits should be taken into consideration and, if possible, expressed in terms of savings when assessing the merit of the investment. The factors that cannot be described readily in monetary terms should be considered in general terms at least.

[a] Improvement of Economic and Social Influence

Robots often take over jobs that are dangerous or that must be carried out in hostile environment. Physically demanding or monotonous jobs can often be carried out by robots alone. Employee training costs or workman's compensation costs alone can sometimes justify robotizing. Eliminating unsafe or undesirable jobs should be a major incentive for robotizing even if accurate cost figures are unavailable.

As industrial robots carry out their work unquestioningly and unfailingly, production control is much easier. Robotizing enables better control by maintaining current production at constant levels thus avoiding production slumps or stockpiles of excessive materials. It also gets rid of troublesome labor management problems by rendering layoffs, short time operation or engaging more workers for short periods to increase production unnecessary. Increase or decrease of production could simply be carried out by adjusting the rate of operation of the industrial robot.

Workers thus benefit by being freed from working in uncongenial conditions or performing boring, repetitive work, and can be placed in charge of the operation or maintenance of the industrial robots in this way advanced to higher levels of employment. If an intelligent robot is, e.g., utilized for wire-bonding of ICs, which is a simple, repetitive job normally done using a microscope and requiring much patience, a worker is no more needed for this arduous operation. Moreover, while a worker may need four to five months of training to become sufficiently proficient at this task the time required for him to acquire the skill needed to operate a robot may be as short as 15 minutes. This means that even newly recruited staff can be easily trained to handle the robots. In Japan, e.g., large numbers of blue collar workers have switched to white collar work as a result of the introduction of industrial robots.

Robotizing provides more employment opportunities, e.g., in the fields of production and operation of industrial robots; it also creates the conditions which allow physically weakened older people or female labor to take up production work. Though the service industry in Japan, e.g., provides most employment opportunities for older people and female labor the manufacturing sector owing to the introduction of industrial robots may also open up employment opportunities for these people.

[b] Quality Improvement

A human worker whose morale is affected through doing boring, tedious work, carrying out a job in an uncongenial, hazardous environment or performing a task that is physically demanding is likely to have the quality of his work suffer. In comparison, a robot may be more consistent on the job and may therefore produce work of a higher quality. The robot is too stupid to mind but smart enough to perform a task better.

[c] Increase in Throughput

When a robot works fast enough to just match the output of the human worker the work will be of higher quality which naturally implies greater net output. There are however often situations whereby a robot can work faster to increase gross output as well. In its own right, this increased throughput is important but improved utilization of capital assets may greatly alleviate the problems of labor shortage and labor productivity.

[d] Greater Production Flexibility

It is possible to have small production and automation runs of medium to large ranges of products, as well as automation of mixed flows of production process. In the past automatic special-purpose machines had been utilized only for large production runs of small ranges of products. Due to the diversification of consumers' demands now about 80% of products made by machines are those with production runs of small to medium quantities. Robotizing is therefore the most suitable way to cope with the present forms of production.

Model changes can be easily implemented in the case of industrial robots. For an automatic special-purpose machine there is a need to replace or remodel the machine when a product change is called for, resulting in costs and time being borne by the enterprise concerned, whereas for the case of the robot only the program needs to be changed when there is a product or model change, which is a very great advantage. With the life cycles of various products tending to become shorter and shorter industrial robots should be the right choice.

[e] Better Utilization of Existing Plant Equipment

Cost reduction in the production system is achieved by the curtailment of plant and equipment investments, reducing wastage and decreasing the proportion of inferior goods, et al. For instance, the proportion of inferior goods could be reduced by stabilized continuous operation, and, with a general tendency for over-thick painting when painting work is carried out by workers the utilization of industrial robots for painting jobs would eliminate this problem resulting in 20% to 30% of the paint being saved. Though there is some initial cost involved in introducing robots in the long term it is worth it. As robots are made light-weight and small with the progress of technical development their efficiency in terms of factory space utilized should be increased. Many factories have carried out large scale installation of robots in order to enhance efficiency. Greater utilization of manufacturing space is highly important as land prices are normally very high. It is hence important to make more effective use of factory space by introducing industrial robots.

[f] Upgrading of Technical Innovations

Technological innovations of industrial robots will expedite the appearance of new industries. For instance, control systems of robots with artificial intelligence can be applied to other machines, oceanic development can be sped up by submersible robots and operations of nuclear reactors can be improved by utilizing robots capable of conducting operations and repairs under exposure to radiation thereby making it possible to further develop the nuclear power industry.

Industrial robots have many effective features and applications giving rise to:

[i]   Increase of productivity.
[ii]   Elevation and stabilization of quality of goods.
[iii] Saving of energy.
[iv]   Better production control.
[v]   Relief from uncongenial working conditions.
[vi]   Greater employment opportunities.
[vii] Elimination of the problem of lack of skilled workers.
[viii] Greater practical use of the capability of workers.

[ix]   Development of new industries.

Industrial robots are now accepted throughout the world. Although a totally unmanned factory may not have appeared yet except as a pipe dream by a researcher of production techniques, the flexible manufacturing system (FMS), which comprises of NC machines, robots, conveyors, and an automated warehouse and computer, has made an appearance. The introduction and spread of industrial robots besides bringing about innovations in production techniques may lead to great social impact.

PROJECT APPRAISAL BY THE PAYBACK METHOD

The simplest form of project appraisal is the payback computation, which provides answers to the questions "How much is it going to cost?" and "How soon shall we recover the investment?"

The accountants and financial advisers will in normal circumstances be looking most favorably on the projects which pay for themselves in a relatively short time. The time needed, which is usually measured in years, is known as the PAYBACK PERIOD. Much depends on the type of industry and the nature of the project but most accountants would have no problem at all in approving proposals which yield a payback period of one or two years, and even possibly three or four years.

Simple Payback

What follows is a simple payback example. A robot replaces a human operator whose wages and fringe benefits amount to $12 per hour at a work-station where 250 days are worked in a full calendar year. The robot would only cost $1.30 per hour to run and maintain. Capital investment for the robot and its accessories is $55,000. The company, which normally operates one eight-hour shift per day, has the option of increasing this to two shifts per day when necessary.

The payback period is computed as follows:-

Simple payback formula:

$$P = I/L - E$$

where:

P = payback period in years
I = total capital investment in robot and accessories
L = annual labor costs displaced by the robot
E = annual expense of maintaining the robot

In this example:

I = \$55,000
L = rate of \$12 per hour, including fringe benefits
E = rate of \$1.30 per hour

There are 250 working days per year, with either one or two eight-hour shifts.

Case 1: Single Shift Operation

$P = 55,000/[12(250 \times 8) - 1.3(250 \times 8)]$

  = 2.57 years

Case 2: Two Shift Operation

$P = 55,000/[12(250 \times 18) - 1.3(250 \times 16)]$

  = 1.29 years

For Case 1 above, the payback period of 2.57 years would probably satisfy the accountants. For Case 2, wherein two eight-hour shifts are justified by sales forecasts and production plans, the much shorter payback period of only 1.29 years would be much more attractive to the accountants.

This represents the least complicated method for evaluating the viability of a new project in financial terms. If the payback period is very short there is positive incentive for proceeding with the project, in which case it is unnecessary to utilize a more complicated appraisal method.

Production Rate Impact on Payback
A robot not only replaces a man at the work-station it might work faster or slower than the man, being sometimes capable of being included in an automatic system which allows more efficient operation of one, two or even more pieces of expensive equipment. When the production rate project appraisal method is utilized these factors will be taken into consideration, these representing the value of capital equipment in the application and the production rate coefficient compared with the manual worker standard.

Let us say that the total capital value of associated machinery and equipment is \$200,000 and the depreciation is 15% of the operating cost budget, giving an annual depreciation of \$30,000 which is represented by Z in the complex formula below. Production rate variations are represented by a production rate coefficient, q, which is the rate by which robotized production either exceeds or lags that which is achieved by a human worker. For example, a rate of 20% above the human rate and a rate of 20% below the human rate would provide values for q of plus and minus 0.2 respectively.

The actual calculations are carried out as follows:-

[a] Production rate payback formula when q = -ve:

$$P = 1/[L - E - q(L + Z)]$$

[b] Production rate payback formula when q = +ve:

$$P = 1/[L - E + q(L + Z)]$$

where:

P = payback period in years
I = total capital investment in robot and accessories
L = total annual labor saving
E = annual expense of robot upkeep
Z = annual depreciation costs of associated equipment
q = production rate coefficient

In this example:

I = $55,000.
L is derived from a labor rate of $12 per hour including fringe benefits, taken over 250 working days in a full year, each day comprising of either one or two eight-hour shifts.
E is derived from a rate of $1.30 per hour, covering the same period as L.
Z = $30,000 (being 15% of the total capital cost of $200,000 paid for associated machinery and equipment).
q is either 20% faster or 20% slower compared with a human worker.

Case 1[a]: Single Shift Operation where the Robot is 20% Slower than the Human Worker

$$P = 55,000/[24,000 - 2,600 - 0.2(24,000 + 30,000)]$$

= 5.19 years

Case 1[b]: Single Shift Operation where the Robot is 20% Faster than the Human Worker

P = 55,000/[24,000 - 2,600 + 0.2(24,000 + 30,000)]

= 1.71 years

Case 2[a]: Double Shift Operation where the Robot is 20% Slower than the Human Worker

P = 55,000/[48,000 - 5,200 - 0.2(48,000 + 30,000)]

= 2.02 years

Case 2[b]: Double Shift Operation where the Robot is 20% Faster than the Human Worker

P = 55,000/[48,000 - 5,200 + 0.2(48,000 + 30,000)]

= 0.94 years

The above calculations should be compared with those for the same project using the simple payback formula. The inclusion of capital utilization and production factors for the work-station in this example has a great effect on the results. That is, though the production rate impact on payback method is more complex than the simple payback method the inclusion of production rates and associated capital utilization in the former can greatly reduce or extend the expected payback period. If the robot is reliable and has no inherent bugs, the shortened payback period should apply to most robot applications, with the robot operating continuously without rest periods and with dependable repeatability. It can be seen from the examples above that for an one-shift operation, the payback period of 2.57 years for the example for the simple payback method becomes only 1.71 years for the example for the production rate impact on payback method, while for a two-shift operation, the payback period of 1.29 years for the example for the simple payback method becomes only 0.94 years for the example for the production rate impact on payback method.

4 BASICS OF ROBOTICS

ROBOT EVOLUTION

Industrial robots have evolved from "hard automation" which can be described as equipment which is designed to accomplish repetitive production-line tasks at a quick pace. This device comprises mainly of drive mechanisms controlled by devices such as timers, switches, cams and mechanical and electrical stops. The function of hard automation is usually very specific, seldom requiring alteration to its operation. The changes needed are often only mechanical adjustments. Due to this, the equipment usually cannot respond to a change in the manufacturing process except perhaps to automatically shut down.

Robots have evolved from hard automation in a number of stages. When speed controllers, sequencers, timers, adjustable stops and other devices were added to these hard automations, robots were born. The name "robot" was chosen because the devices had a small degree of flexibility.

The addition of servo controls, which are feedback devices that generate position signals, represents the next stage in robot evolution. The position signals would be compared to the original input positioning signals when the robot moved and when these two signals coincided the robot would cease movement.

The adaptation of computer control was the next and the greatest advancement. The capability to react to externally generated signals was introduced resulting in the robot giving the impression that it possessed human intelligence. The cost of the robot also grew as it became more and more sophisticated. This deluxe, general-purpose robot which was equipped with the latest in computer control technology and capable of all kinds of motions was an over-kill in many applications. This gave rise to the special-purpose robots which still have a high degree of sophistication but less capability and cost less than the general-purpose robots. The rapid development of the microprocessor for robot control made this change possible. Low cost microprocessor controllers have made these special-purpose robots a still greater asset to the manufacturing and industrial organizations.

MECHANICAL UNIT

The mechanical unit is a system of mechanical linkages and joints which can be moved in various directions to carry out the work of the robot and may be called the robot arm in some instances.

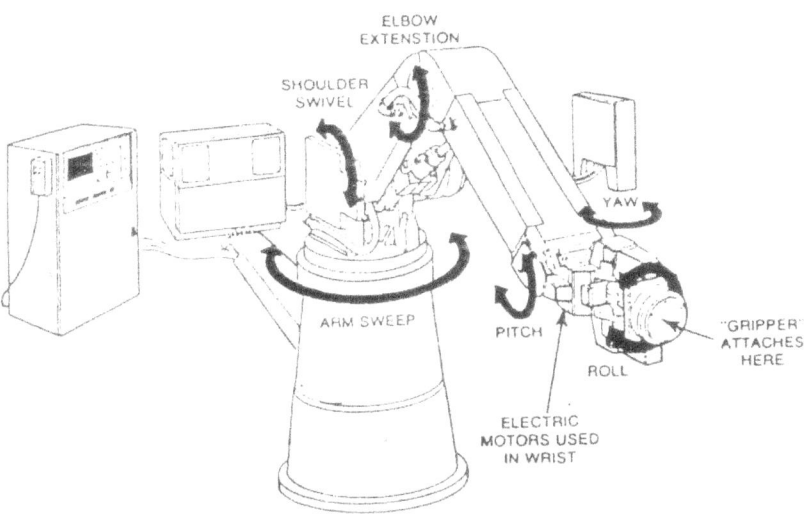

ACTUATORS

Actuators are drive mechanisms which position the manipulator to a predetermined point. They can be pneumatic or hydraulic rotary motors, electric motors, or pneumatic or hydraulic cylinders.

Some robots are positioned employing a combination of these methods. The mechanical actuation of very basic robots may be carried out through bell cranks and cams. Mechanically actuated robots are often classified as automation.

The concept of industrial robots appears to have been established first in the patented "programmed article transfer" by George C. Devol of the United States in 1954. In 1958 an American company, Consolidated Control Inc., developed a digital-controlled robot. In 1962 the first models of Unimates (Universal automate) and Versatrans (Versatile transfer) which have been the prototypes of robots used in the greatest numbers were produced by Unimation and AMF respectively. The USA and Japan are amongst the leading users of industrial robots today.

SOME TERMS AND DEFINTIONS

There are various kinds of industrial robots, with the basic terms associated with each of them being similar. The following are the common terms:-

DEGREES OF FREEDOM

The number of degrees of freedom a robot has represents the number of intricate motions and the complexity of the task the robot can perform. There are generally three main degrees of freedom, or axes, adopted by the various types of robot coordinate systems. They are called the base and arm axes.

There are also usually three additional axes at the extremity of the robot arm, which are a unit commonly called a "wrist", whose main significance is the ability to orientate the gripper or any other end-of-arm tooling. Some or all of these six degrees of freedom are found in most robots. A seventh axis is added if the whole robot is placed on a traversing slide.

The following are included in the wrist axes:-

[a] "Roll" - rotation in a plane which is perpendicular to the end of the arm.
[b] "Pitch" - rotation in a vertical plane through the arm.
[c] "Yaw" - rotation in a horizontal plane through the arm.

Pneumatics are in general most suitable for fast movements and light payloads, which are two characteristics of limited-sequence robots. Hydraulic actuation provides the ability to position and hold heavy loads for long periods without slip. Electric drives are on the other hand considered to be the simplest to control.

Most limited-sequence robots are pneumatically actuated whereby one major advantage over hydraulic actuation is cost. Pneumatically operated systems can be supported by factory air while hydraulic operated systems usually need a separate hydraulic power source for each individual unit.

GRIPPER
A gripper is a hand-like device that is sometimes referred to as the "end effector". It holds the tool that does the work or handles the material which is being worked on. Most grippers are simple open/close devices which are actuated mechanically, pneumatically, hydraulically or with electric motors. There is a virtually unlimited variety of grippers which can be adapted for robot use.

CONTROLLER
The controller is the brains and nervous system of the robot. It controls the movements of the robot, stores position and sequence data and interfaces the robot with the "outside world". A controller can be any programmable device from a simple mechanical revolving drum switch to a complex electronic computer.

POWER SUPPLY
The power supply provides and regulates the energy, which can be electrical, hydraulic, electro-hydraulic, pneumatic or even mechanical, for the robot's actuators.

NON-SERVO ROBOTS
Non-servo robots are robots which are restricted to three or four degrees of freedom. To limit the amount of travel, mechanical stops (end stops) are used on each axis. There are usually only two positions for each axis to assume - left/right, up/down, extend/retract.

SERVO-CONTROLLED ROBOTS
Servo-controlled robots are robots whose axes of motion are programmed to allow the robots to move and stop anywhere within their limits of travel rather than only at the extremes. They can be divided into two classes, namely, point-to-point and continuous path.

PAYLOAD
The payload is the maximum load which a mechanical unit can position. This load capacity is measured at the gripper. The specifications of many robot manufacturers give the capacity under static (stationary) conditions, which is not necessarily the load the robot arm can handle at maximum reach and speed. The maximum load capacity of the robot depends not only on the strength of the arm but the velocity at which

the load is to move and how fast the load is accelerated and decelerated.

SPEED

The cycle-time of a robot, which is the amount of time needed to carry out one complete sequence of operations, is determined by the speed at which the mechanical unit travels. In determining the speed, the payload and strength of the mechanical unit are major factors.

ACCURACY

Accuracy is the placement repeatability of the robot, i.e., how close the robot can position its mechanical unit to a given point. It is the robot's ability to return its mechanical unit to the same given point when repeatedly cycled. For each robot with its payload, cycle frequency and work capability accuracy is important.

Small pick-and-place robots can achieve a much greater degree of accuracy than larger robots, e.g., accuracies of plus/minus 0.1 mm are achievable by some low-technology robots while placement and repeatability from plus/minus 0.25 mm to plus/minus 2 mm is common.

INTERFACING

For every application the robot has to interact with something in the execution of its programmed tasks. Unless a part is available for the robot to handle and the robot has been signaled that the part is present, even a simple part transfer operation cannot be successfully accomplished. When the robot is interfaced with other related equipment there will be transmission of information in two directions.

Interfacing can be carried out through simple on/off signals by means of electrical or pneumatic contacts, or, through more complex electronic signals. The lines over which the robot receives signals are known as inputs while the lines over which it sends signals to the external equipment are outputs.

DEFINITION OF ROBOT

A definition for robots from the Japan Industrial Robot Association (JIRA) is "mechanical systems capable of performing movements resembling those of the upper limbs of a human being or having sensing and recognition capacity with the capability of controlling its own behaviour".

Another definition for a robot, from the Robot Institute of America, is "A robot is a reprogrammable, multifunctional manipulator designed to move material, parts, tools or specialized devices, through variable programmed motions for the performance of a variety of tasks.".

A feature which a device must have in order to be considered a robot is to be able to operate automatically on its own. That is, there has to be inbuilt intelligence, a programmable memory or simply an arrangement of adjustable mechanisms which command manipulation.

COORDINATE SYSTEMS

Cartesian Coordinate Robot (or Rectangular Coordinate Robot)
This configuration comprises of only straight axes of motion, wherein the horizontal arm of the robot moves in and out and its carriage moves up and down on a vertical column which is able to move transversely in a straight-line motion along its base.

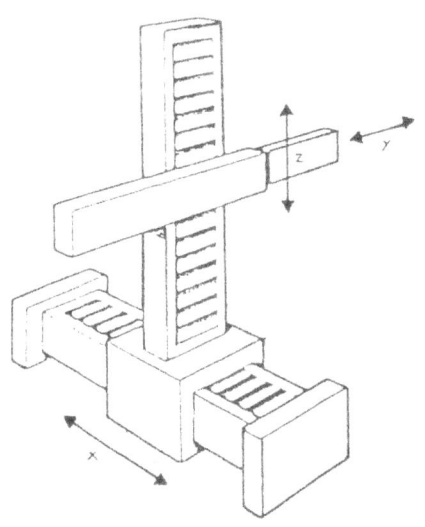

CARTESIAN COORDINATES

Cylindrical Coordinate Robot

This configuration comprises of a horizontal arm mounted on a vertical column which is in turn mounted on a rotating base; the horizontal arm of the robot moves in and out and its carriage moves up and down on a vertical column, these two members rotating as a limit on the base. The working area or envelope is a portion of a cylinder.

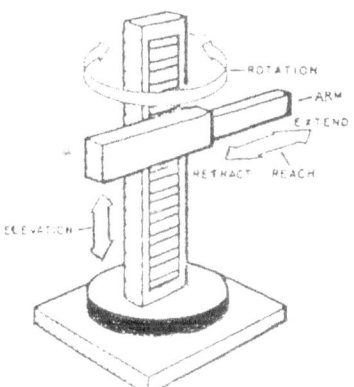

CYLINDRICAL COORDINATES

Spherical Coordinate Robot (or Polar Coordinate Robot)

This configuration is the same as the turret of a tank, with an arm moving in and out, pivoting in a vertical plane and rotating in a horizontal plane about the base, the work envelope being a portion of a sphere.

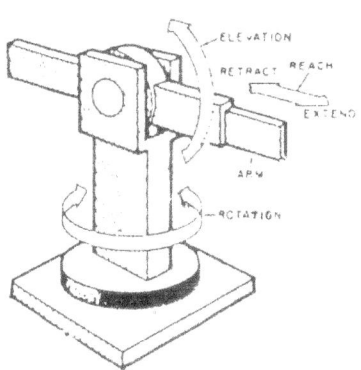

SPHERICAL COORDINATES

Jointed-Arm Robot (Articulated or Revolute Coordinate Robot)

This configuration comprises of a base or trunk and an upper arm and forearm that move in a vertical plane through the trunk. Located between the upper arm and the trunk is an "elbow" joint. At the shoulder joint there is also rotary motion in a horizontal plane. The working envelope is more or less a portion of a sphere.

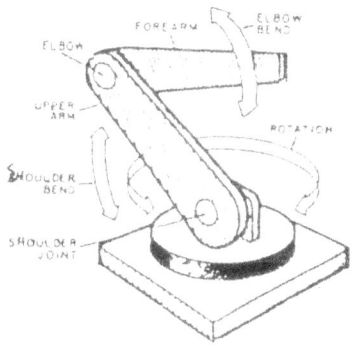

JOINTED-ARM COORDINATES

CLASSIFICATION OF INDUSTRIAL ROBOTS

[i] Manual Manipulator
A manual manipulator is a machine that can be operated directly by a person.

[ii] Sequence Robot
A sequence robot is a manipulator with working steps operating in accordance with pre-set information on sequential order, condition and position. Sequence robots are usually of two types, namely, fixed sequence robots and variable sequence robots. Fixed sequence robots are robots whose pre-set information cannot be easily changed while in the case of variable sequence robots the pre-set information can be easily changed.

[iii] Playback Robot
This is a robot which is first taught a certain working procedure, then memorizes the procedure and can continuously repeat its operation.

[iv] N. C. Robot
This is a manipulator that operates according to numerically coded information.

[v] Intelligent Robot
An intelligent robot is one that can perform various functions itself through its sensing and recognizing capabilities.

PROGRAMMING METHODS
Most robots operate with an open-loop program whereby they continuously repeat the same set of motions without modifications. Robots can be programmed in one of the following ways:-

Manual Programming
The programs of simple robots are created and changed by physically fixing stops, inserting punched cards, setting limit switches, arranging wires, or, in the case of air-logic units, fitting air tubes. Most of the robots that have two to four axes of motion and relatively few steps in their program are called limited sequence robots.

Leadthrough Programming
The memory utilized for the leadthrough programming is similar to that for the walkthrough programming. In leadthrough programming the operator drives the robot through the sequence - with a control console or a separate control box the robot is slowly moved through the program wherein at the

end of each motion the action is entered into the robot's memory.

Walkthrough Programming

In the case of more complex robots wherein magnetic tapes or discs or a computer memory generate the robot commands, the program can be "taught" by walking the robot through the operating sequence. The operator manually moves the hand and arm of the robot and then signals the memory to record the motion or location. Control handles are utilized by the programmer to walk through the program. Such robots are normally designed specifically for painting work. In this kind of robots there is one control for the arm motion and another for the painting sequence.

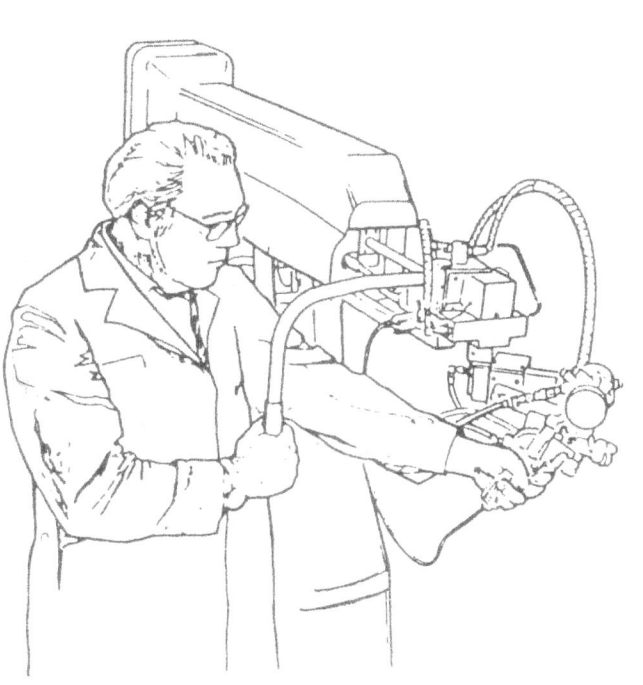

5 END EFFECTORS DESIGN ACCORDING TO WORKPIECE ORIENTATION AND CONFIGURATION

INTRODUCTION

End effectors are the moving parts that have to grasp, lift and manipulate work-pieces without causing any damage to them and without letting go of them. The hands of the robot are clumsy travesties compared to human hands, having fewer articulations and without any sense of feeling or touch. However, they can be designed to withstand high temperatures, which enables them to work with parts that are red hot. They are more suitable for handling objects with sharp edges, covered with corrosive substances or simply too heavy for human hands to grasp. There are many ways for a robot to tackle a job, depending to a large degree on the nature of the material being handled. The options for end effectors include the following:-

[i]   Mechanical grippers
[ii]   Hooks
[iii]   Thin platform or spatula
[iv]   Scoop or ladle
[v]   Electromagnet
[vi]   Vacuum cups
[vii]   Sticky fingers (using adhesives)
[viii] Quick disconnect bayonet sockets

CHOICE OF GRIPPING METHOD

The design of the gripper and the choice of gripping method are influenced by the following factors:-

[i]   Geometry and weight of the work-piece
[ii]   Shape, size and position of the gripping surfaces
[iii] Shape permanence and change during manufacturing
[iv]   Demands for accurate positioning
[v]   Accessibility for gripping and requirements for orientation in the working position and when processing
[vi]   Working environment

Standard Hand

This hand is inexpensive and all-purpose and will accept a virtually infinite variety of customized fingers, fingers which can be tailored according to the parts, which should be of moderate weight, to be manipulated or moved. The finger action and the force needed to grip an object sufficiently tightly are provided by simple linkages, with the fingers exerting their maximum clamping force on the part at the completion of finger closure.

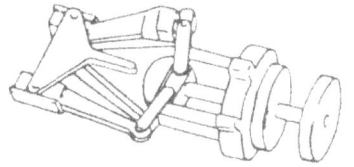

Fingers with Self-Aligning Pads

For ensuring a secure grip on a flat-sided part self-aligning pads for fingers are of value. With the employment of these pads misalignment of the part is highly unlikely.

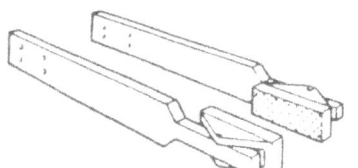

Fingers for Grasping Parts of Different Sizes
A finger design does not have to be restricted to parts with a limited range of sizes, e.g., the fingers can be equipped with extended pads which have several cavities for parts of varying sizes and shapes or for parts which change shape during processing. The industrial robot can be pre-programmed to position the hand such that the proper cavity will match the location of the part.

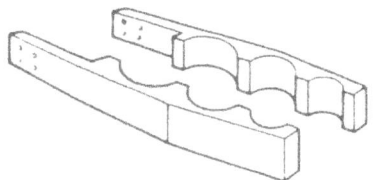

Cam-Operated Hand

With the cam-operated hand heavy or bulky objects can be handled easily. The cam-operated hand, which is more expensive than the standard hand, is designed to hold the part such that its center of gravity is very close to the "wrist" of the hand. The twisting tendency of a heavy or bulky object held by the hand is minimized by the short distance between the center of gravity and the "wrist". There is however a sacrifice for achieving this "close coupling" of hand and part, namely, a specific cam-operated hand design will accommodate only a very limited range of object sizes.

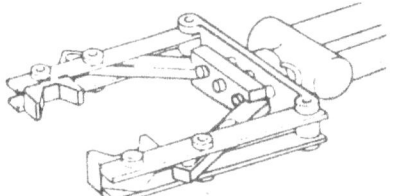

Wide-Opening Hand

A wide-opening hand may be utilized when the part to be picked up is not always in constant orientation or at the same side. This hand will sweep the inexactly placed part into its grasp as it closes.

The wide-opening hand can shorten the time required to reach for the part if the part to be grasped is precisely positioned for pick-up, the hand traveling the shortest path to the part and skipping the extra step of making its final approach to the part from one specific direction. It is suitable for parts whose weight is moderate. When opened the hand develops a low force and when closed the force is at its maximum.

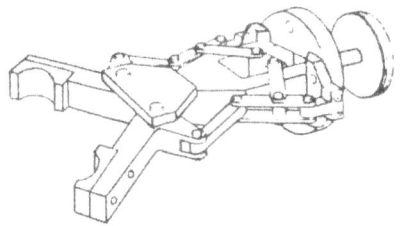

Cam-Operated Hand with Inside and Outside Jaws

This special hand is suitable when a part is re-orientated prior to its removal. The hand can grasp the part on the outer diameter (OD) by utilizing the outer self-aligning pads when the part is orientated as shown in the diagram below. The inner pads will grasp the inner diameter (ID) if the part is turned over.

When the grasped surface of a part is changed greatly between the time it is placed in a machine and the time it is removed the same principle applies. To deal with most changes in OD, ID or other dimensions a special hand can be designed.

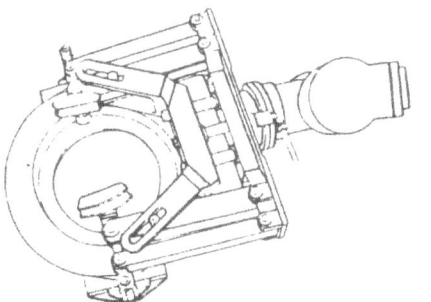

Special Hand with One Movable Jaw

This is one of the most economical hands due to the simplicity of its design. When there is any access underneath a part, e.g., when it is on a rack, a hand with single-acting jaw can be utilized.

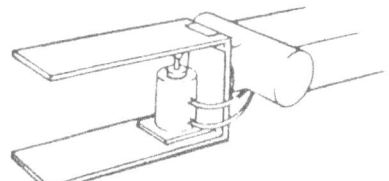

Special Hand for Cartons

This is a dual-jaw hand which will open wide to grasp inexactly placed light-weight objects, e.g., it can be utilized for the lifting and placement of cardboard cartons. Actuators and jaws can be re-mounted in any of several positions on a fixed back plate, which makes it practical for the same dual-jaw hand to move small cartons on one day and larger cartons the next day.

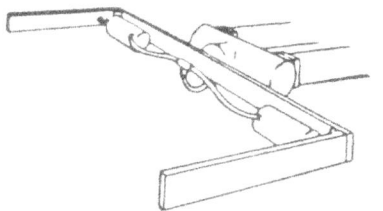

Special Hand with Modular Gripper

With a pair of pneumatic actuators, this special hand is one of the many special hand designs for industrial robots. It is suitable for parts which are light-weight. Lifting capacity is dependent on the friction developed by the fingers but heavier parts could be handled if the fingers could secure a better hold, e.g., under a flange or lip.

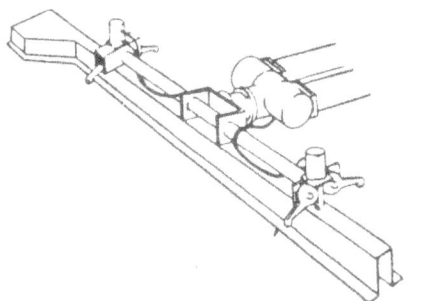

Special Hand for Glass Tubes

The forte of this special hand is secure grasping of relatively short tubes. Even when tube length varies to some degree pick-up will still be equally effective. The fingers of the hand close in two stages - first traveling through an arc until they are vertical, and then the actuator draws them together axially - linear movement in the second stage of closure is selected to accommodate the range of tube lengths to be handled.

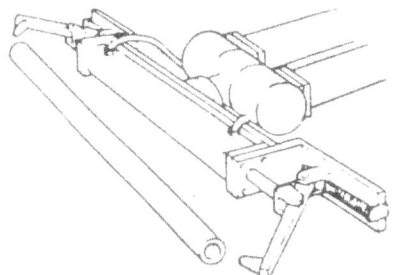

Special Hand - Chuck Type

This is a relatively simple mechanism comprising of three fingers and a single actuator. It is suitable for handling drums and similar large cylindrical parts. By means of a chain and sprockets the actuator drives all three fingers simultaneously, the fingers expanding against the inside diameter of the drum. This hand is capable of picking up drums of various diameters.

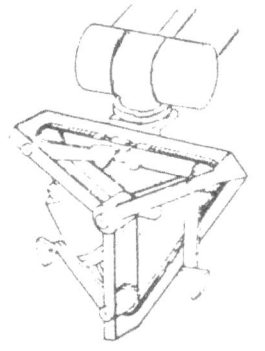

Double Hand

A double hand with double actuators can be utilized for a robot application which requires the hand to remove a finished part from a machine and replace it with an unfinished part. For example, the hand will pick a part out of the chuck of a machine, swivel and place a new part back in the chuck. The industrial robot with this hand therefore does not have to expand time to put one object down before it manipulates another - the hand seldom makes a trip while empty. When the double hand is utilized the parts should not be of more than moderate weight.

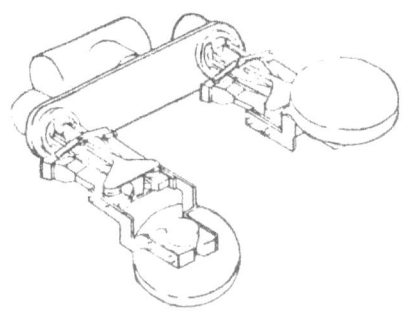

Vacuum Cup Hand

Vacuum pick-up has the advantage of the magnetic pick-up. It is much less susceptible to work-piece side slip. The vacuum cup hand is often an excellent choice for light to moderate-weight glass, plastic, ferrous and non-ferrous parts.

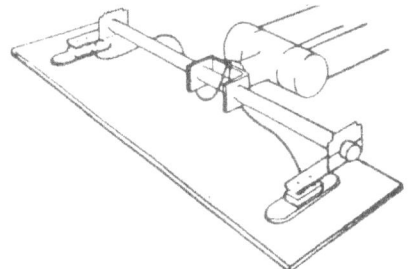

Simple Vacuum Cup Hand

A simple vacuum pick-up can easily handle fragile parts such as cathode ray tube face plates. It is more reliable than the magnetic pick-up. It has well-designed telescoping vacuum lines for long-reach arms.

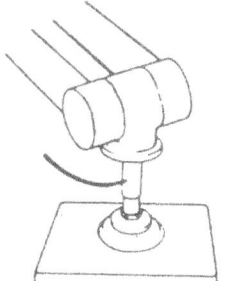

Expansion Bladder Hand

An expandable bladder in the form of a cuff, which is supported by a rigid back-up ring, will be suitable for handling large cylindrical vessels with flexible walls which are difficult for a mechanical hand and fingers to grasp. In the diagram below the illustrated plastic container with tapered walls is a typical part which the bladder can easily handle. However, a given bladder design will only be able to handle one size of vessel. There are two options for the bladder hand, namely, an internally expanding (in ID) bladder as is shown in the diagram below or an externally expanding (in OD) bladder after insertion into a vessel. A suitable alternative for this application can be the vacuum pick-up.

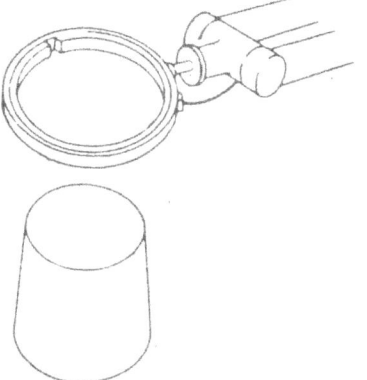

Electromagnetic Pick-Up

The electromagnetic pick-up is good for use on flat surfaces such as ferrous sheets or plates and can handle objects of varying sizes whose weight should not be more than moderate in order that side slippage is avoided. Precise positioning of the part to be picked up is unnecessary and grasping is instantaneous, both of which save time.

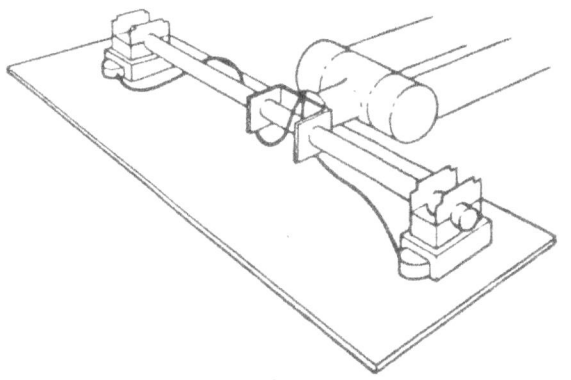

Inert Gas Arc Welding Torch

Arc welding with a robot-held torch is another application of industrial robots. While the welds can be single-pass or multiple-pass the most effective use is for welding simple-curved and compound-curved joints as well as carrying out multiple short welds at different angles and on various planes.

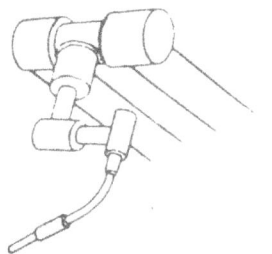

Ladle

Industrial robots are suitable for hot and hazardous jobs such as ladling hot materials, e.g., molten metal. The robot can be programmed to scoop up and transfer the molten metal from the pot to the mold and then carry out the pouring, for such applications as piston casting and permanent mold die casting. Dipping techniques will often prevent dross from forming in the mold.

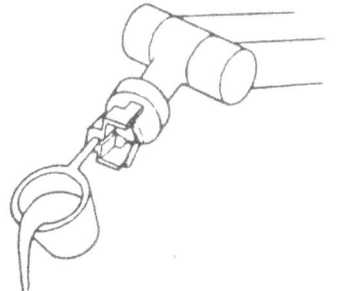

Spotwelding Gun

A series of spot welds on flat, simple-curved or compound-curved surfaces can be carried out by a general purpose industrial robot maneuvering and operating a spotwelding gun.

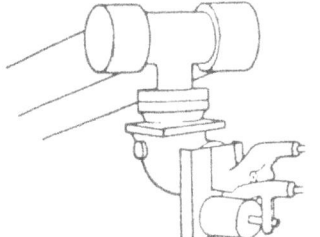

Stud-Welding Head

An industrial robot can be equipped with a stud-welding head. Studs are fed to the head from a tubular feeder which is suspended overhead.

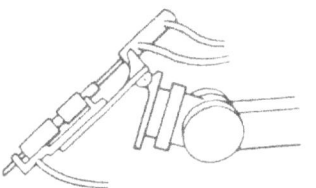

Heating Torch

A heating torch can be manipulated by an industrial robot to bake out foundry molds by playing the torch over the surface, letting the flame stay longer where more heat input is needed. Because heat is applied directly fuel is saved, and, the bake-out is faster than it would otherwise be if the molds were conveyed through a gas-fired oven.

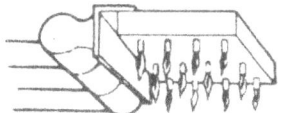

Pneumatic Nut-Runners, Drills and Impact Wrenches

For carrying out nut-running and similar operations in hazardous environments general purpose industrial robots are especially suitable. They can also be used for drilling and countersinking operations, with the aid of a positioning guide. Mechanical guides increase the locating accuracy of the robot and help shorten positioning time.

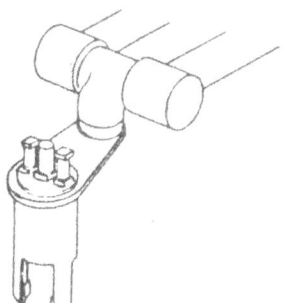

Routers, Sanders and Grinders

A routing head, grinder, belt sander or disk sender can be easily mounted on the wrist of an industrial robot wherein the robot can rout work-piece edges, remove flash from plastic parts and carry out rough snagging of castings.

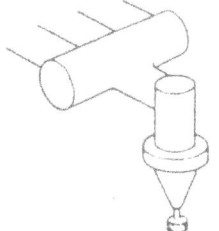

Spraying Gun
The ability of the industrial robot to carry out multi-pass spraying with controlled velocity makes it suitable for automated application of primers, paints and ceramic or glass frits, as well as the application of masking agents before plating. The industrial robot is often a better choice than a special-purpose setup which requires a lengthy changeover procedure for each different part, for short or medium length production runs. Moreover, the robot can spray parts with multiple surfaces and compound curvatures.

6 ROBOTIC INTELLIGENCE

Cybernetics, which comes from the Greek word meaning "steersman", is the science of control and communication in both machines and living organisms and was invented by Norbert Weiner. It is especially concerned with things which are self-controlling or adaptive. Due to changes in its environment, an adaptive system alters its behavior, e.g., an automatic pilot which is used in aeroplanes alters the course of the plane because of changes in the wind speed.

Artificial Intelligence
Artificial intelligence (AI), which is about making machines do intelligent things, is a field that is closely related to cybernetics. Though machines need to be able to "think" to do something intelligent experts disagree about what this means. However, some who believe that a machine which "learns" from past experience or responds to things happening to it can be called a "thinking" machine. Others argue that thinking machines must have feelings and want to do things. This implies that a "thinking" robot, e.g., would have to want to pack boxes because it enjoyed this work.

Clever Machines
Due to their ability to simulate or mimic intelligent human activity such as the processing of visual information and speech through ingenious programming, computers are the cleverest machines available. Computers can be utilized to control other machines, such as robots, to make them behave "intelligently".

"Intelligent" Computer
A computer can be programmed in two basic ways. Algorithmic programs, which are often used for robots, work by evaluating all the possible alternatives in a situation. Heuristic programs, often AI programs, are "cleverer" as they take short-cuts to arrive at decisions by remembering from past experience the best way to solve a problem, e.g., a chess-playing robot computer could work out the best moves by being provided the rules of the game.

Speech Recognition
Computer programs have been developed to provide robots the ability to recognize spoken commands using a microphone as an electronic "ear". The average person knows thousands of words but a computer would need a massive memory to understand even a tiny fraction of them. Besides this the computer has to take into account the different ways people speak. Programming the computer to recognize only a short list of words spoken by one person, which are needed for the robot's job, is much simpler.

Vision
Increasingly robots are equipped with machine vision that allows them to "see" and behave

"intelligently". The intelligent part of this is not the robot, the computer or the TV camera eye, but the computer program, which analyzes and interprets what the "eye" sees - something which is very complicated. Humans are very selective in what they actually see, which is difficult to simulate with a computer.

7 DEVELOPMENTS IN ROBOTICS

Robotics is a fast-moving and exciting subject. There are many robotics research projects going on around the world. There are more and more arm robots being used in factories along with other automatic machines. By utilizing more and better sensors together with clever computer programs for their control, robots are also being made more "intelligent". This includes mobile and other kinds of robot "servants" in the factories. Robots are also becoming cheaper, e.g., a micro-robot costs about the same as some home computers.

DEVELOPMENTS

How Computers Recognize Words
Each word produces wave-like patterns of sound which are converted by a microphone into electricity, the waves varying according to the different sounds in a word.

The height of the wave, which is an electrical voltage, is gauged many times a second. These measurements are recorded as a sequence of numbers. They are then turned into a digital code of "pulse" and "no pulse" bits. The computer can then use these bits to identify the word.

How Robots Recognize Things
To recognize one or more objects a machine vision system can be programmed. A machine vision system can recognize an object in a pile and tell a robot how to pick it up correctly for packing in a box.

The vision system focuses on one part of the pile. It projects stripes of light over it to judge its distance. This information is dispatched to the robot's computer which can work out the outline of an object from the breaks in the stripes of light. As the computer is programmed only to identify the outline of the object it will not recognize anything else.

The computer can work out the position of the object in the pile by comparing this outline with views of the object stored in its memory. This information is then dispatched to the robot in the form of instructions to its motors.

The computer manipulates the robot to pick up the object without damaging it or any of the other objects. The robot turns the object the right way round for packing. This sequence is repeated for all the objects in the pile.

Industrial Mobile Robot
This is a driverless forklift truck used in an automated warehouse or factory, which has an on-board computer and power supply and uses sensors to navigate.

Nuclear Reactor Robot
This is an arm robot which is designed to be used in the core of a nuclear reactor. The arm is suspended from a long hollow chain with the control cables for the arm passing through the chain.

Robot Servant
This robot can be programmed to perform tasks such as serving drinks at a party and speaking to guests with its synthesized voice. Such robots can also be made to do housework.

Robot Metro
In Lille, France, a completely automatic robot train has been built. The trains are computer-controlled and can switch between tracks and stop automatically at stations.

Robot Computer Assistant
This is an arm robot that goes up and down in a honeycomb storage cell to find special cartridges that contain computer data and delivers them to the computer and replaces them after use.

Two-Armed Robot
This robot is designed to work alongside humans on a production line. Its arms enable it to perform complex assembly work. It can perform two tasks simultaneously. Its base contains control microcomputers.

Walking Robot
This is a four-legged robot which walks and can climb stairs, built by Japanese scientists. Other researchers have been trying out six-legged and eight-legged designs that walk like insects.

Modular Arm Robot
Some arm robots are made in modules or small units, e.g., base, arm, wrist, et al., which can be combined in different ways to make a robot suitable for a particular task.

Robot Cleaning Machine
A free-roving, industrial floor-scrubbing robot with navigation sensors and probably a sensor to detect when the water becomes dirty has been developed.

Android

Androids, which are robots that look and act like humans, have been made and are used mostly for exhibitions and shop displays. They are powered by electric motors and hydraulic pistons.

8 GUIDELINES FOR AUTOMATION

The following is a useful guide/checklist for automation:-

Technical Considerations
[1] Evaluating the suitability of an installation/estimating of profitability.
[2] Conduct of technical analysis.
[3] Selection of robot and peripheral equipment.
[4] Tooling requirements.
[5] Safety requirements.
[6] Provision of utilities.

Implementation Considerations
[1] Layout and interface.
[2] Development of grippers and tooling.
[3] Installation and start-up.
[4] Human relations.
[5] Training of staff.
[6] Trouble-shooting and monitoring of performance (maintain record of downtime).
[7] Maintenance.
[8] Possible improvements.
[9] Documentation (technical details, interface, et al. - for staff who resigns a replacement has to be recruited).

Objectives of Automation
[1] Cut down manufacturing costs by increasing productivity.
[2] Improve product quality and reliability by reducing or eliminating human errors.
[3] Better utilization of labor, materials and production floor.
[4] Maximum utilization of machines and equipment.
[5] Reduce waste.
[6] Compensate for shortage of skilled or unskilled labor.
[7] Reduce training cost.
[8] Improve safety of workers.

Reasons for Robotic Installations
[1] Better utilization of machines.

[2] Shorter throughput times.
[3] Take over less congenial work.
[4] Good flexibility.
[5] Reduce manual work.
[6] Uniform quality.
[7] Solve labor shortage problems.
[8] Increase efficiency.
[9] Enhance prestige.

Planning Procedures
[1] Form project team.
[2] Define objectives of project.
[3] Identify robotic application areas.
[4] Carry out economic analysis.

Investment Costs
[1] Purchase price of robot.
[2] Cost of gripper.
[3] Cost of automating machine and equipment.
[4] Cost of automatic tool replacement.
[5] Size of magazines.
[6] Cost of fixtures.
[7] Cost of installation of utilities.
[8] Cost of investigation and planning.
[9] Control panel.
[10] Transformer.

Considerations Relating to Production
[1] Machine occupation.
[2] Operating time.
[3] Automatic working cycle.
[4] Automatic feed and extraction of work-pieces.
[5] Simple set-up work.
[6] Complete chain of operations.

Lower Production Costs by Reducing
[1] Labor costs   -   direct

	-	indirect
[2] Operating costs	-	electricity, water, air, et al.
	-	service, maintenance
[3] Capital costs	-	machines, equipment
	-	factory space
	-	work-in-progress
	-	stores
[4] Material costs	-	direct materials
	-	indirect materials for lubrication, cleaning, et al.
[5] Administration costs	-	workshop management
	-	planning and ordering routines
	-	personnel administration
	-	advertisements, mails, et al.

Principles for Grouping of Machines
[1] Flow-orientation.
[2] Functional-orientation.

Good Layout
[1] Maximum flexibility.
[2] Coordination amongst machines.
[3] Maximum utilization of production floor.
[4] Accessibility.
[5] Minimum distance.
[6] Inherent safety.

Technical Assessment
[1] Weight of product.
[2] Positioning of components into fixtures/output components.
[3] Conduct of technical analysis.
[4] Type of peripheral equipment.
[5] Test of grippers.
[6] Planning of layout.

Supplementary Equipment for Automatic Working System
[1] Automatic material feed.
[2] Automatic material handling.
[3] Automatic process control.
[4] Automatic quality control.

Types of Flexibility
[1] Product flexibility.
[2] Method flexibility.
[3] Item flexibility.
[4] Machine and planning flexibility.
[5] Volume flexibility.

Preparatory Work
[1] Arranging the supply of fresh materials.
[2] Handling of finished details.
[3] Readjustment of machines.
[4] Servicing of tools.
[5] Preventive maintenance.
[6] Programming and commissioning of new programs.
[7] Quality control.

Problems affecting Process Reliability
[1] Tool wear.
[2] Tool breakage.
[3] Swarf removal.
[4] Supply of cutting fluid.
[5] Different cutting fluids for different materials.
[6] Variations in dimensions and quality of raw materials.

9 DEVELOPING AND BUILDING A ROBOT

INTRODUCTION
It is not difficult to build a robot nowadays. Parts can be purchased in the market and Do-It-Yourself kits can be bought at the store or through mail order. Fully assembled robots can even be purchased for a reasonably small sum. The hardware and software aspects have to be considered when developing and building a robot.

HARDWARE OF ROBOTS
The hardware of a robot refers to the machines, the units, the mechanisms which are used. Generally it is possible to distinguish between the mechanisms themselves as end items (as actual consumer products) and the parts or components which make up these end products. An industrial robot and a personal robot which are both assembled and ready-to-go are end products while DC negative motors, transistor drives, solenoids, accommodators, feedback transducers, tactile end-effectors, grippers, sensor products, positioning systems, et al., are parts or components which may be used to produce an end product. Let us here take a look at the actual merchandize, the wares, the durable items of commerce which are available in the field of robotics.

ROBOT HARDWARE REQUIREMENTS
A useful, general-purpose, multi-function, intelligent robot should be in possession of the following general capabilities:-

[1] Sensory capabilities such as:
    [a] being able to hear with great acuity
    [b] being able to see
    [c] being able to touch and feel
    [d] being able to taste
    [e] being able to smell

All of the above capabilities should be greater than or as great as human capabilities, e.g., night vision, sensitivity to various impressions which humans are not normally sensitive to (such as carbon monoxide and poisonous gases) and animal level hearing.

[2] "Perceptual" capabilities, i.e., the ability to identify and discriminate objects, which is connected to the ability to pattern and the ability to interpret.
[3] Ability to carry out commands, directions, et al. The robot should be maneuverable and able to

manipulate and handle materials of all shapes and sizes and thus requires a "body" which includes the following:

[a] Arm, wrist and hand maneuverability.
[b] Memory for remembering what to do.
[c] Coordination and agility.
[d] Executability, speed and reliability.

[4] Ability to communicate in terms of the principal mode of human communication/speech.

[5] Ability to think and create.

[6] Probably also emotional, imaginative and intuitive capabilities.

Expecting a robot to be in possession of all the above-mentioned capabilities may seem too idealistic. However, are all these capabilities really necessary? Are robots which possess such capabilities really needed? Are such elaborate mechanisms really needed? Such mechanisms are indeed needed for many purposes, the most important being probably space exploration and home and industry use.

Robots are required for space exploration for a simple reason. Humans on earth could direct the activities of tele-operated machines working in near space. There would be problems once such activities take place much beyond the moon as it takes 2.4 seconds for a command signal to get from the robot to earth and back to the robot on the moon. This is probably too long a time to wait for highly complex and dangerous activities, e.g., 2.4 seconds may be too long for a robot to wait to know what to do next if it were about to step off a cliff.

When the distances are longer the problem would be worsened, e.g., it takes 40 minutes per signal for a robot on Mars and a human operator on earth to communicate. Intelligent machines which can decide on their own what is to be done to solve problems encountered in their environment are therefore necessary for carrying out exploration on the planets, and certainly beyond the solar system. Home and industrial robots are very capable machines that are replacing human labor.

INDUSTRIAL ROBOTS
Industrial robots are utilized, as the name implies, in industrial - mostly manufacturing - situations while personal robots are used by individuals in their homes. There are more and more developments and applications in industrial robotics. The wide-spread use of single-function and multi-function robots on the factory floor has been pioneered by the Japanese, which shows that increased productivity and efficiency

could be achieved by using robots. In 1980, 80,000 industrial robots were at use in Japan while in the US there were at most only 1,500.

Though the US has not used robots as widely as Japan they have been strong in discoveries related to the technology of electro-mechanical robots. There have been 100 or more robot-manufacturing companies in the US. A number of computer-based companies have branched into robotics, e.g., IBM, which has a long-term robotics research program, has created AML (A Manufacturing Language), developed in 1975 for use on the IBM PC. The leading producer of industrial robots in the US is Unimation Inc., which has more than 5,000 robots in use around the world.

Most robots are used in industry, especially manufacturing, e.g., in casting dies, welding, lifting heavy objects, loading machine tools, spray-painting, dangerous tasks, et al. In manufacturing they are widely used for assembly, setting up queues, quality control, et al.

COMPONENTS OF ROBOT MECHANISMS

The attempted development and use of industrial robots which can use sensors to respond to an environment have been rather limited. Industrial robots are provided with different sorts of maneuverability depending on the main task at hand, generally programmed to carry out a specific job at any time and may be reprogrammable and have multi-tasking units. Sensor technology has been rather limited and undistinguished. Video cameras, proximity sensors and tactile sensors have been used in helping a robot to position itself to carry out a task and to help in the actual performance of those tasks., e.g., infra-red sensors have been used to sense the presence or location of objects which are to be placed on pellets.

There has also been rather limited development in the perceptual capabilities and pattern recognition capability of robots, with AI work having been carried out in this area. Sensors with great acuity and software which can be used to filter and pattern sensations have to be developed. Sensors which can help a robot to detect the difference between standing still and being in motion and recognize objects belonging to either of these two classes are available. A robot capable of such things as face recognition, multi-part recognition and sortability, and so on is of course desirable, e.g., a robot-computer which could "see" one part set on an undifferentiated background part set.

A robot's maneuverability and versatility are dependent on the adequacy of its power/drive system and its ability to "do" something. Technically, it is end-effectors which enable a robot to "do" things. To be able to position itself to do things a robot has to be versatile in what they do and highly maneuverable.

Air (pneumatics), liquid (hydraulics) or electricity (motors) is used by the power/drive mechanisms of robots. Other forces, e.g., solar energy, may be used if feasible. The load and the accuracy with which movements are to be effected will determine what power/drive units are used.

Pneumatic drive units are unsuitable when precision is needed as air is difficult to control, but they are cheap, light and, though not very strong, quick. Electric motors are suitable when accurate movements are required and when small power units and light loads are expected though they are most expensive.

Hydraulic methods are used when there are heavy loads and when gross accuracy is required. A really versatile robot should have all these three methods available for use at any one time.

MANIPULAORS

Besides the power/drive aspect of a robot, a tool which is able to do things is necessary. This tool is one which is isomorphic to a human arm and wrist in most if not all cases. The power/drive unit enables the robot to do something. The robot's arm-wrist does it.

The most common end-effector is a grip mechanism known as a manipulator, which has multi-degrees of freedom, an open-loop chain or series of mechanical joints and linkages that are driven by actuators and that is capable of moving an object from point to point along prescribed trajectories.

A really useful robot can be expected to have six degrees of freedom, being capable of moving in six directions - forward, backward, left, right, up and down. Many industrial robots however have five or less degrees of freedom.

An actuator is a motor that causes motion in response to a signal and has joints which are like finger joints and links that are joined together in a series of mutually responsive and interacting motions.

Movement takes the form of a change of position or rotation around a point. It makes use of a device that moves relative to a point of references and a goal or working point, the kind of motion to take place having to be specified. For instance, a gripper is to move from point A to point B. Its trajectory of motion might be one of the following: straight line, circular path or rotation.

A manipulator is thus able to rotate or change position relative to a point of reference, using one of the three trajectories, moving in some combination of the degrees of freedom, in order to move from point A to point B which are specified in terms of the coordinates x, y and z.

INDUSTRIAL ROBOTS IN THE MARKET

Unimation is a leading robot maker whose president Joseph Engelberger is one of the industry's leaders and a most articulate spokesman on the future of robots. One of the robots Unimation makes is the PUMA robot, which is driven by electrical motors that are necessary for performing precision work and has joints which are each controlled by a micro-processor. General Motors was one of the first to use this robot on its shop floors.

Another robot used on the shop floor for assembly work is IBM's Rs-1, which can be controlled through the IBM PC through using the AML programming and robot control language. The Intelledex Model 605 is not just a mechanical arm, unlike the above two robots. It has high precision, being able to repeat to within 0.001 of an inch motions which it has been taught. This level of precision and even higher precision are needed for electronic assembly work. With an integrated vision system the robot can find its way to within 0.002 of an inch. It is programmable in BASIC and is able to set itself up on the factory floor. It is also able to handle a variety of weights and tasks.

Most industrial robots used today are however single-function. They have a certain form and are bolted to the floor.

PERSONAL ROBOTS
Kits or fully assembled personal robots which are general-purpose, programmable, maneuverable and quite capable are available. These robots have wide applications. Provided the user has the patience and knowledge, these robots are really quite useful in the home, at school and at work or play, though the software for these mostly experimental robots is somewhat limited there being possibly still some bugs in them.

COMPONENTS OF THE PERSONAL ROBOT
Most personal robots have sensory capabilities which enable them to respond to the external environment. With the proper software they are able to map out "spaces", e.g., a room, and to "learn" their way around in those spaces by using photodiodes, sonar transducers, tactile and obstruction sensors, motion sensors, light sensors and sound sensors. For the robot to do these things effectively, instructions and various kinds of sensors are required. A personal robot can carry out any number of useful tasks by being able to detect light and darkness, the presence and absence of sound, and the presence and absence of obstructions.

The personal robot can be programmed to carry out various tasks, e.g., waking up household occupants at certain times, turning the lights in a house on and off depending on the conditions and working during the day and "sleeping" at night, tasks which depend on its being able to detect light and darkness.

Human voice can actuate a robot which is able to detect the presence and absence of sound. This kind of robot may respond in speech to voices that it hears, entertain by singing, teach, serve as a watch-dog, guard or sentry, answer the door and welcome guests, et al.

The sensors of personal robots, as is in the case of industrial robots, though quite remarkable are generally low-level sensors, which can be improved with advancement in sensor technology.

With suitable sensors the personal robot can detect still and moving objects, which is a form of perception, a form of differentiating the mass of sensations or pattern recognition. The sensors also enable the robot to determine the range of objects within its sensory environment.

Sensors of great power and versatility are available for space exploration and military applications, e.g., a military radar network which can "see" and track an object the size of a grape-fruit from a distance of over 30,000 miles can cost hundreds of millions of dollars. With declassification by the military and mass uses found for these sensor devices, their cost should decrease dramatically.

We should bear in mind that to work sensors have to be of the right type and connected to the right software or interpretation hardware with the interface being at the right point. The human being is capable of many complex functions, e.g., making judgment, solving difficult problems and identifying people and objects, many of which are not easy for the robot to duplicate, which poses a great challenge to AI researchers.

Though the invention of chips enabled thousands of transistors to be compressed onto a surface a fraction of the size of a transistor the size of the chip could still be further reduced. Cramming enough computing power into a reasonable amount of space is a problem, which makes it very difficult to duplicate the power of the neurons in the human brain. Without this improved technology, a robot that could sense and perceive with the power of a human brain would have to be huge, the size of a football field, which would not be of much help. With the development of atomic and organic chips it could be possible to duplicate the neurons in the human brain in some manageable amount of space, e.g., the size of a human head. To duplicate or isomorph the sensory and perceptual capabilities of a human being ingenuity is indeed needed.

Many personal robots can speak, e.g., with the aid of phoneme speech synthesizers which are available in several models. A robot can speak in any of the major languages without difficulty though "recognizing" and understanding such speech when spoken to may still be a problem.

The electric motor and omni-directional wheel-triads system are the power/drive systems mostly used in personal robots. By utilizing batteries robots can work without direct and constant connection to an external power source, and, if rechargeable batteries are utilized they can detect when their batteries are running low and in some cases tell a present human of their plight and/or find an electrical outlet and plug themselves in for recharging. A robot with the capability of utilizing solar energy may not have this external dependency and if a suitable mechanism is utilized for locomotion may be able to move around level surfaces, hop over obstructions, climb down ropes, go up ladders, et al.

An arm-wrist, which is sometimes a standard feature of personal robots, is occasionally used. It enables the robot to pick up objects of a size and weight that depends on the strength of the gripper mechanism (approximately one or two pounds, though much more powerful grippers are available). In one model the arm-wrist tool is capable of rotating up to 250 degrees and can rotate the wrist by 180 degrees.

Robots should ideally have two arms as it is evidently far easier to perform many tasks with two hands than with one. In fact two-armed personal robots have been available in the market-place.

It is a dream to have a "creative" robot. With the realization of atomic and organic "chips" or electronic circuits that can duplicate human neurons this is feasible.

PERSONAL ROBOTS IN THE MARKET
The earliest personal robot had the form of a turtle or mouse that could be used experimentally. Since the early 1980s the turtle and related mechanisms such as the mouse have been around. There are a number of personal robots that are mass-produced and generally available in the market, e.g., the Hero-1, which was the first commercial home robot, introduced in January 1983 by the Heath Company, the RB5X introduced by RB Robot Corporation, the TOPO and B.O.B./xA of Androbot Corporation, the Genus of Robotics International, the Marvin Mark 1 of Iowa Precision Robotics Inc. and the HUBOT of Hubotics Inc.

These personal robots can be programmed in a number of languages, e.g., LISP, FORTH and HEX (machine-level code), and can perform many tasks, e.g., "hearing" and responding to sounds, speaking, answering the phone, vacuum-cleaning, walking, singing, playing games and reading. There are many companies that produce components for constructing robots for the home or work-place. There are motors of different powers available for the drive mechanisms, solenoids and other switches, various arm-gripper mechanisms, sensors available as sensing devices and a whole range of computer equipment for developing "intelligent" and programmable robots. Most home robots come in pre-assembled form as the market for robot hobbyists is quite limited.

FUTURE ROBOTIC MECHANISMS
The development of robotic technology depends on the entrepreneurial instincts of robotic engineers and pioneers and on how people are generally responsive to the availability of personal and industrial robots.

The US and Japan are the two closest rivals in robotic technology. The utilization of robots in Japan has brought about high productivity gains. In the US retooling using robotic tools has been a major drive though the capital requirements of retooling are very high.

As the market for industrial robots is small and selective, manufacturers of industrial robots could afford to develop special-purpose, limited-task robots. There has been development of versatile industrial robots that can be used in many different industries, e.g., flexible manufacturing systems (FMS).

Personal robots which are multi-purpose, versatile, adaptable and economic will change our personal and social life in unimaginable ways, e.g., robots that can act as personal secretary, house-keeper, researcher, companion for children and animals, general overseer, scheduler and general worker.

A robotic side-industry that engages in the development of attachments and appliances that robots are able to use, e.g., ovens, washers, driers, vacuum-cleaners and television sets, may be viable.

For work in dangerous and hostile environments robots can take over from human beings.

Agricultural robots which can distinguish between weeds and edible plants and hence can serve in checking weeds, take soil samples, pick fruits and grade fruits will be very useful.

Other areas of robotic development also include robots involved in prison work and police work.

ROBOTIC SOFTWARE
Robotic software are programs, similar to programs that run on computers; these software are malleable and reprogrammable wares. To do work, robots need software. All these programs serve one main purpose, namely, specify the set of procedures or algorithms for carrying out a task.

Through the use of a programming language programs are coded into a machine. Some of the programming languages used in robotics are just elaborations of already existing languages for programming computers while other languages are new developments, e.g., TINY BASIC, which is a subset of BASIC, is used by RB5X and LISP, which is used widely in AI work, is also used as a robot operating language. IBM has on the other hand developed a robot programming language called AML (A Manufacturing Language). This language has been considered by IBM to be the most advanced robot programming and computer programming language around.

There are at least three levels of software being utilized. There is first the application software that is designed to do something, e.g., telling a robot how to open a door, directing it how to cook a chicken and directing it in the performance of daily chores. This is software that directs a robot to do something which we want done such as clearing the garbage, playing with the children, et al.

At a higher level is the "control" software, which directs the robot on how it is to inter-relate all of its different parts, how it is to inter-relate its voice with its action, movements, hearing, vision, et al. This software is analogous to the brain where control is exerted over these movements, while the application software is analogous to a person moving his hands and legs.

At a still higher level are the various meta-control rules that have been proposed for robotics, Asimov's three laws of robotics being among the most well-known. These laws are aimed at governing the relationship between the "control" software, the application software, the hardware that effect the required actions and society. This level of software is analogous to the ideas that are produced by the cerebral cortex, which is the front portion of the brain, the center of thought where all the "higher" functions like language, thought, imagination, et al., take place. Asimov's three laws of robotics are as follows:-

[1] No robot may injure a human being directly, nor may it allow a human being to be injured because of a failure to intervene directly.
[2] Robots must obey human orders except where it conflicts with the first rule.
[3] A robot must preserve itself, except where such preservation would result in violations of laws (1) and (2).

These three laws however will not be found to be adequate when subjected to analysis.

The following is a diagram of the three types of robotic software:-

Meta-Control Software

Program In Memory   **Control Software**

Application Software

Sensors   **End-Effectors**

Environment In Which Programs Are Executed

CLEVER ROBOTS

Robots have been given artificial intelligence. It is necessary for clever robots to have computers with clever programs in them. It very difficult to program in computers some things which seem easy for us, e.g., a computer does not always know what a word means.

Most computers and robots require very careful programming as there is no room for mistakes. If something happens that it does not expect a computer cannot guess what to do. For robots to be able to work out more details for themselves artificial intelligence has to be improved.

Thus, the importance of software development.

10 THE FUTURE OF ROBOTS

Introduction
Robots have played an increasingly important role in our lives. The robotic revolution has already started in any case. The utilization of robots brings many advantages, e.g., robots are efficient and labor-saving, do not complain or tire out and can do many of the things that we cannot or do not want to do.

America's Futuristic Plan
President Ronald Reagan in an address carried by all three television networks in March 1983 caught his audience (and some of his advisers) by surprise when he unveiled a futuristic plan for nullifying the destructive power of the USSR's nuclear-tipped intercontinental ballistic missiles. He made a call for the construction of hundreds of "intelligent" robot battle stations to form a multi-layered defense in outer space to shield the United States and her allies against sneak attacks. He also entreated his military researchers to provide the technological impetus to back up his concept for ensuring the survival of the American way of life. There were however some negative reactions to Reagan's novel approach to national defense.

Such a proposal might initially seem like an entirely impractical pipe dream. What the President actually did was asking America's engineers to provide a powerful fleet of orbiting battle stations that sweep repeatedly across the sky, always ready at a moment's notice to shoot down hundreds or thousands of enemy missiles inflight, each tracing out a different trajectory and each swooping down toward a different terrestrial target at a speed of about 15,000 miles per hour, roughly ten times the muzzle velocity of a 0.22 rifle bullet. An attack, if it occurs at all, will likely involve intense salvos of military missiles launched in waves purposely concentrated in time and place to overwhelm the defense. The defenders thus have to be extraordinarily agile, having a few seconds only at most, to eliminate one missile before attacking another. In space the defenders will have to be capable of rapidly making intelligent decisions when faced with situations they have never faced before. There is also likely to be a great use of laser weapons in space warfare.

For years robots have been used in space to handle a range of useful assignments including simulated military attacks though the thought of forming a protective barrier around the country with a swarm of orbiting robots might seem somewhat far-fetched. Most space-faring military robots are however much friendlier than that. America's reconnaissance satellites have been taking snapshots of strategic regions of the earth for years. A satellite automatically ejects a film packet that releases a parachute on entering the atmosphere and is snatched from the air by a special military air-craft when it finishes taking a particular series of pictures.

Robots in Space Exploration

The US has got their Surveyor space-craft to dig a narrow slit trench in the lunar soil under the direction of technicians stationed on earth. In the 1970s, the Russians had similarly used two compact moon carts, the Lunohods, that wheeled around the lunar land-scape making scientific measurements and taking photographs that were transmitted back to Russian laboratories. Another impressive space-age robot was the Viking space-craft sent by NASA to the surface of Mars with the view of achieving new scientific insights and finding new life forms on the planet. The Viking worked in the infra-red and not the visible part of the electromagnetic spectrum with its two "eyes" and was equipped with a simple arm that could push rocks out of the way, dig holes and pick up Martian soil samples under the watchful eyes of its designers back on earth. It had a separate little "finger" that measured wind speed and direction and was also equipped to measure Mars quakes and to find living microbes if any were hidden in shallow burrows under the planet's surface. Viking's findings were relayed back to scientists at the Jet Propulsion Laboratory, whose scientists sent back detailed instructions to control the space-craft's behavior when necessary, by radio transmitters whose energy was supplied by nuclear isotopes.

A large manipulator arm was created in Canada for use on the Shuttle orbiter. This manipulator arm was directed by the astronauts onboard the Shuttle instead of an operator located on the ground. Being 50 feet in length it is probably the largest remote manipulator ever built, having been designed for use in the weightlessness of space. It was not strong enough to support its own weight if positioned on the ground.

Robots and remote manipulators have been enormously successful in space exploration. It has been the opinion of some experts that most of the work done in space is better carried out by robots than man as there are definite advantages, e.g., the booster rockets needed to transport robots into space could be much smaller and lighter, complicated life-support systems are not necessary, routine expendable supplies such as air, water, tooth-paste and shaving cream would be little needed and problems encountered by certain sensitive onboard equipment and interference to measuring and observation work which is caused by the bodily movements and breathing of the human being would be avoided.

The development of a new free flying robot, which would be especially useful for missions wherein the space-craft had to rendezvous with a disabled satellite or one that requires fresh fuel supplies, for use in connection with the Space Shuttle had been proposed by Grumman Aerospace. The robot could also enable the space-craft to handle such a mission on its own without outside help. However, the disadvantage would be that its rendezvous rockets spew out a sooty gas which coats and contaminates delicate space-craft components like thermal sensors and solar cells. The construction of a remote tele-operator called POV (Proximity Operations Vehicle) especially designed to minimize space-craft contamination had been proposed by the Grumman engineers. A 30 cubic-foot box, the POV had a gangling arm protruding from each of its four corners, each arm carrying four thrusters that are powered by non-contaminating nitrogen gas. The POV was maneuvered by remote control by the astronauts aboard the Shuttle when it flew out to meet a satellite. It could be provided a picture from either of the two television cameras aboard the craft, one pointing forward in the direction of travel, the other pointing off, by its control panel. A grappling device would emerge from its front end, get hold of the disabled satellite and take it in tow when it reaches its destination.

Equipped with sets of grappling arms, later versions of the POV could repair satellites without bringing them back to the Shuttle. Similar tele-operators could be used in assembling large structures in space and in the maintenance and repair of satellites in hard-to-reach orbits, including those at the geosynchronous altitude.

Robots in Ocean Exploration
Remote manipulators operating on the bottoms of the deepest oceans might resemble the robots already in use or planned for use in outer space in many respects. These could be utilized in rescue and salvage operations and the recovery of minerals and hydrocarbons in large quantities. For example, robots could be utilized to mine manganese nodules that are scattered across huge expanses of the ocean floor, which could be the basis of a major new industry. Several large companies have in fact been trying to perfect submersible robots to find and gather these sea-floor nodules.

The Glomar Explorer is one such robot, which uses a sled-like nodule plucker that propels itself forward with horizontally mounted spinning screws and has two long metal cylinders running along the sides of the crawler and a spiral blade resembling the thread of an Archimedes' screw curling gently around each cylinder. The crawler rakes up any loose nodules that have been spotted by its video camera as it slithers along the bottom of the ocean floor. The nodules collected are then crushed and mixed with salt-water to make a slurry which could be pumped to the surface through an 18-inch pipe that is more than three miles long. The crawler also has a television camera, high-intensity lights and a sonar unit which is used in positioning the crawler relative to a network of bottom-mounted sonar transducers. A box-like station that is used to pump the slurry to the surface hovers above the crawler, 100 feet above the bottom of the sea. The crawler and the free-floating pumping station move in tandem at a speed of about one knot.

Robots could also be used for undersea research, e.g., a remote controlled robot capable of remaining submerged for 30 days or more. Such a robot could be in the form of a free-wheeling robot that carries another smaller robot just in case the parent craft spots something that requires a more detailed examination. The robot could be towed by a ship over the mid-oceanic ridge and would be able to explore hundreds of miles of the sea-floor each month rather than just several square miles.

Robots for Dangerous Jobs
Robots could be utilized in clearing radioactive spills, handling toxic wastes and dangerous mining jobs. For example, Odetics Corporation in Anaheim, California, has developed such a robot called the Functionoid, which is an ungainly insect-like, six-legged robot which could skitter over rough terrain and lift a 3,000-pound load and is controlled by ten on-board computers. Robots could also be put to military uses such as clearing hazardous mine-fields, chemical warfare and reconnaissance.

Robots for the Physically Handicapped
Much of the similar clever technologies that permit robots to work in such hostile environments as mine-fields, the bottom of the ocean and outer space have been used in the creation of a new breed of manipulators to help the physically handicapped, people who have to spend their lives in very difficult circumstances. For instance, movements can be triggered and controlled by electrical impulses which flow from a micro-computer through multi-colored wires to a number of electrodes taped over the major muscles of a handicapped person's incapacitated legs. The micro-computer with the help of a program figures out the exact time to send each impulse into the leg muscles of the handicapped person. The movements and responses of the ankles, knees and hips of the handicapped person will be monitored by sensors and this information will be fed back into the computer. With advanced technology this device can be a small microprocessor that is implanted pace-maker style in the person.

Robots in the Field of Athletics
With the use of robots novel and interesting approaches to athletic training for specialists and non-specialists alike are possible, e.g., a person can enroll in a special program which is designed to help him learn how to execute a perfect tennis stroke. The robot will be the trainee's instructor and the trainee will "wear" the robot while he is under instruction. Immediately after arriving at the training center dressed in his tennis outfit, a computer will scan the trainee's body with its electronic eyes to measure the lengths of the various segments of his arms and legs so that he can be fitted into a robot-like device which resembles a whole-body prosthesis. In order to make the robot very comfortable to wear, every of its mechanical linkage will be tailored for a snug fit. On being worn by the trainee the robot instructor will become a very efficient tutor urging the trainee with supreme patience towards perfection in the game. The robot's eyes will analyze the ball's trajectory and the trainee's body position to determine exactly how the trainee should hit the ball as the ball crosses the net. Using its voice-synthesizer chip the robot will make helpful comments on his game and will understand his replies though it may have problem knowing how to respond to emotional outbursts. The robot will show him how to execute the various standard back-hand and fore-hand strokes over and over again, gradually correcting any flaws in his form, once the trainee is on the tennis court. For instance, if the trainee tends to face the net too much the robot will gently but firmly turn his feet towards the side-lines and if he is hitting the ball too late the robot will make adjustments in his form a little at a time so that he will be hitting the ball at the proper location in front of

his body soon. Whenever he makes any important changes in his game the robot will confide in him vocally. Every shot the robot teaches the trainee will be optimized by its computer for his own particular body size and shape. The robot is able to repeat the same perfect strokes over and over for him until the "feel" of them is forever burnt into his brain.

The robot tennis instructor will have some interesting characteristics, e.g., he will allow the trainee to hit the ball the same way the trainee's favorite player does, taking into consideration physiological differences. The trainee just tells the robot whom he admires and which shot he wants to execute and the robot will search his electronic memory for the desired characteristics. The robot will then play the stroke through the trainee's body as many times as the trainee likes in quick succession. The trainee can also imitate the styles of more than one star tennis player. Whether the trainee's favorite player is left-handed while the trainee hits right-handed or if his favorite player is female while he is male, or vice versa, it will not make a difference to the robot whose computer will make the necessary adjustments automatically.

Perhaps one day there will be new types of sports events wherein the competitors are allowed to wear their robots during play if robotic sports training becomes popular and profitable.

Recreational Use of Space-Age Robots
Robots may make it possible for one to go on special vacation "trips" and do the standard things enjoyed by tourists such as wandering through the Sistine Chapel without ever leaving one's favorite arm-chair and taking the elevators to the top of the Eiffel Tower. With this "virtual reality" one could even "picnic" in the North Pole or "climb" Mount Everest without any fear of personal injury at all.

Medical Robots
Robots can assist in surgery, e.g., a surgeon could perform delicate micro-surgery by using his own hands to control the movements of a much smaller robot, for instance, one of microscopic size, whose movements could be easily scaled appropriately with a computer. With proper feedback from sub-miniaturized sensors, the surgeon could "feel" the textures of various tissues and organs in order to operate successfully on detached retinas, damaged nerves or clogged arteries without invading the body except to make a small opening to insert his tiny manually controlled robot. This kind of surgery would be much safer and less painful than the conventional type; in fact, so little extraneous tissues would be damaged that it is possible for the patient to return home on the same day.

Conclusion
Many kinds of robots are featured in science fiction, e.g., androids or humanoids, which resemble human beings. In the movie "Star Wars" there was R2D2, an extremely intelligent robot, a robot that was perhaps too human and intelligent. There is perhaps the constant fear in us human beings, whether conscious or

unconscious, that robots may become too clever by half and pose a threat.

AI researchers have in fact been developing robots and computers that are as similar to human beings as can be. The Japanese have been developing a new super breed of computers which contain knowledge, think and make decisions better than any human being, the Fifth Generation computers. They have also been into related projects, e.g., super computers and robots capable of reproducing themselves.

Robots may one day end up doing all the work while people will not have to work at all. On the other hand, many fear losing their jobs to robots. Hopefully there will not be a repetition of the First Industrial Revolution whence the Luddites had gathered together and gone on a machine-destroying mission as machines had become a threat to their jobs and livelihood.

One day a robot or computer may be so intelligent that it will have a life of its own. The moot question is whether a robot or computer can have consciousness just like we human beings. What produces our consciousness and can consciousness be bestowed on machines? Some scientists seem to agree that it is possible for machines to have consciousness; they seem to believe that if the human being is duplicated as precisely as possible this may be a reality. A number of books on computer consciousness have already been written.

Is it conceivable that one day a new breed of super robots will rule the world while we human beings will become their slaves or servants? This may seem like pure science fiction to some. There is however really nothing much to fear about robots. All we need to do is to disconnect their power plugs if they were to become too threatening. However, the robot is meant to make our life better.

11 APPENDIX: VARIETIES OF ROBOTS AND MISCELLANY

Japan's Fifth Generation Computer Systems
The Japanese government had unveiled plans for a highly ambitious program to design and build a completely new range of computers for the 1990s which is called the Fifth Generation Computer Systems. As projected, these machines will be capable of understanding natural language and speech, interpret the visual world, tap large knowledge bases and solve problems through deductive and inductive inference.

Though these plans were certainly ambitious, even grandiose, they were not unrealistic in the light of the country's industrial success in other fields and would give the US and Europe a run for their money.

TRANSMISSION OF WRIST MOTIONS

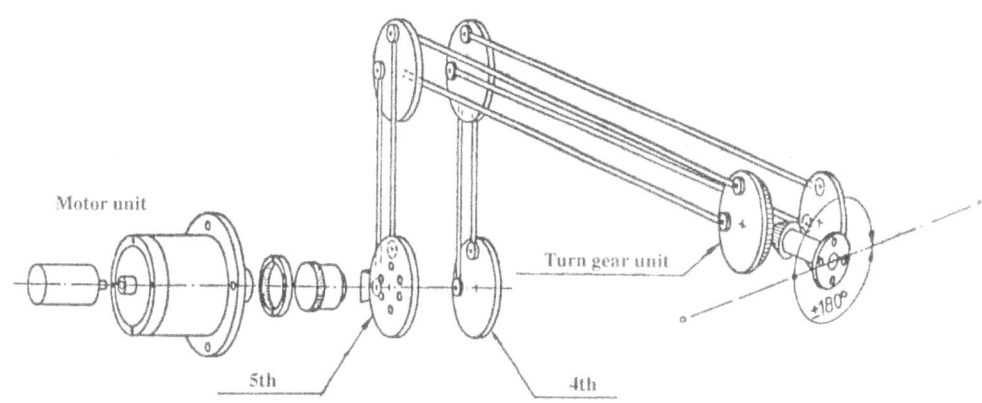

Motor unit

5th

4th

Turn gear unit

+180°

Robot calibration and timing belt adjustment are important routine maintenance procedures.

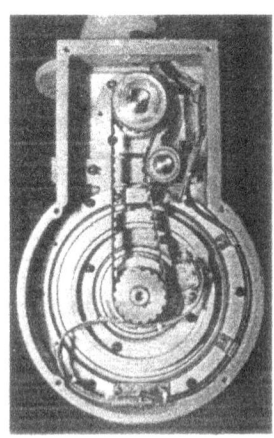

The IRB 9000 Robot - Pendular Robot for Fast, Flexible Spot Welding
The IRB 9000 robot produced by ASEA Robotics, Inc. provides speed and flexibility in spot welding applications. It has a pendular configuration and six axes. It can weld difficult-to-reach points and offers a typical cycle time of 1.4 seconds per spot weld.

The robot has a repeatability of +1.0 mm, can handle payloads of up to 90 kg and can be installed on the floor, mounted on a wall or column or suspended in an inverted position from an overhead structure.

The robot employs ASEA's standard S2 robot controller and programming is simplified by means of ASEA's ARLA high-level, dialog programming language. Commands are entered with a portable reach pendent. Robot movements are programmed with a joystick, which reduces programming time by up to 25%.

A picture of the IRB 9000 robot is presented below:-

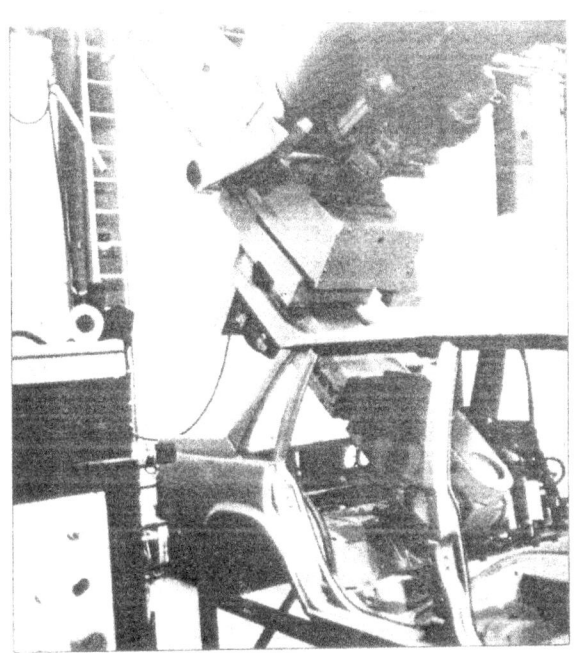

A Robot that Performs Welding Work

Close control of parts, fixture or positioner, and weld gun are necessary for automated arc welding. A robot performing such welding work is shown in the picture below:-

A Robot that Serves as a Butler - Omnibot

He glides into your room to wake you with your favorite music in the morning. He is the perfect butler at your house parties serving food and drinks to guests. He can walk the dog, change the sheets, dust the shelves or deliver a poem.

He responds to your commands without any complicated computer programming or accessories. You just record the robot's program into a cassette tape and Omnibot will execute it.

Omnibot can store up to seven different programs, which it performs any time or day one may specify during a seven-day cycle. Via a remote control console it can respond and a built-in microphone allows one to talk over the on-board speaker. With a joystick one can send Omnibot anywhere in the house.

Below is a picture of Omnibot:-

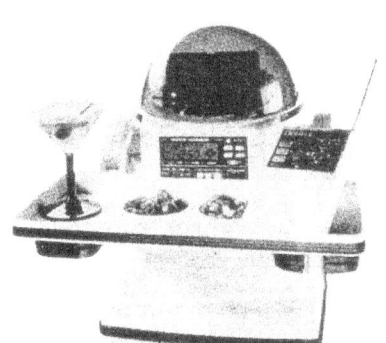

Surveillance Robots

Surveillance robots depend on good vision, hearing and communications. To obtain clear images, Surveyor (A) transmits what it sees through a black video over two video channels via a conical microwave antenna. RRV (B), which had been used in the Three Mile Island incident, was submersed in a dark containment-building basement and required strong adjustable headlights to illuminate contaminated walls so that the television camera could film them.

The picture of a surveillance robot is presented below:-

Assembly Robots

Robots can access almost any location normally not accessible to a human. The robot in the picture below is using a DC brushless electric motor with a microprocessor-based monitoring and control system. Adapters allow quick changeover of fastening tools of various sizes to provide maximum flexibility. Such an application in the "repair bay" of an assembly system could remove and replace defective fasteners and signal that the operation has been completed successfully.

The picture of an assembly robot is presented below:-

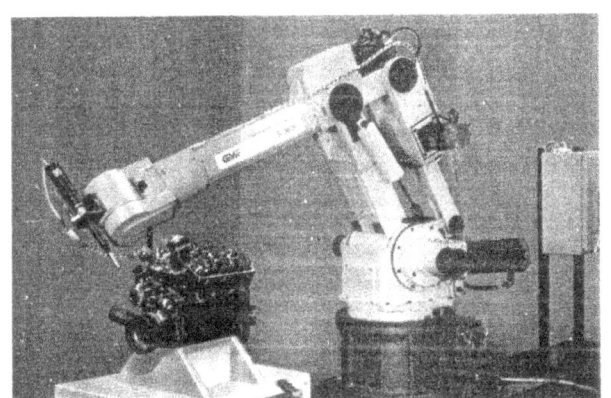

Industrial Robots

Industrial robots are frequently used when flexibility is needed and the parts to be handled are not too heavy. In the picture below a robot is loading a machined light-metal casting into a locating fixture from another line. At the other end of this extensive APS system completely assembled automotive transaxles would emerge.

The picture of an industrial robot is presented below:-

An Independent Robot

In the robotic workcell for applying masking to printed circuit boards shown below no operators are required - the robot manipulates the boards under the dispensing nozzle - production is doubled:-

124

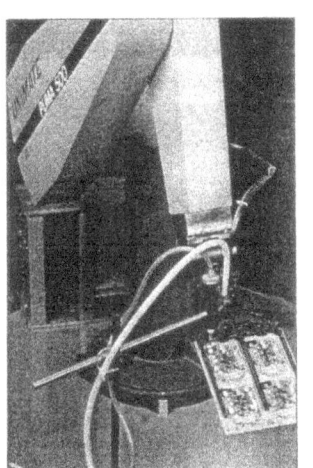

A Commercial Robot

The extensive end-of-arm devices, which include a three-dimensional system and a gripper which has a servo-controlled grip opening and grip force, developed by the NBS were added to the commercial robot shown in the picture below:-

A robot that walks on stairs:-

A six-legged computer

Inspection Robots

The problems of ultrasonically testing composite aircraft parts have been alleviated by the precise assistance of robots.

The picture below provides an overall view of the H80 robotic inspection system testing for defects in a composite aircraft part:-

The Spine Robot

The unique Spine robot with its snake-like movements would charm robot buyers. Automakers are likely to take to the robot. The Spine design is one of the most unusual robot configurations in the market.

The picture of the Spine robot is presented below:-

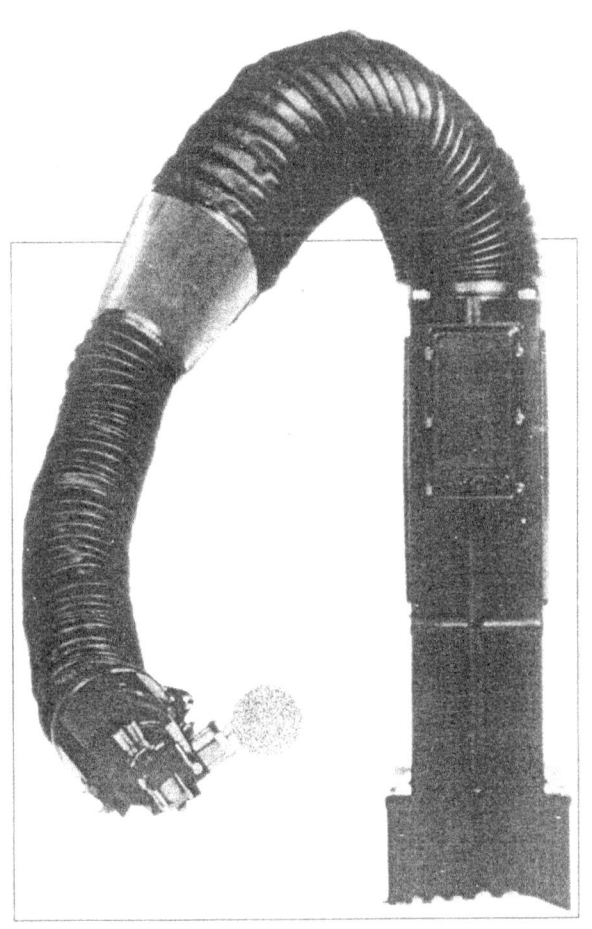

Construction of the Spine Robot

The Spine robot comprises of two sections, each section consisting of a series of disks held together by four hydraulically actuated cables (shown in inset of picture below). The large work envelope of the Spine robot provides great flexibility.

Below is the picture showing the construction of the Spine robot:-

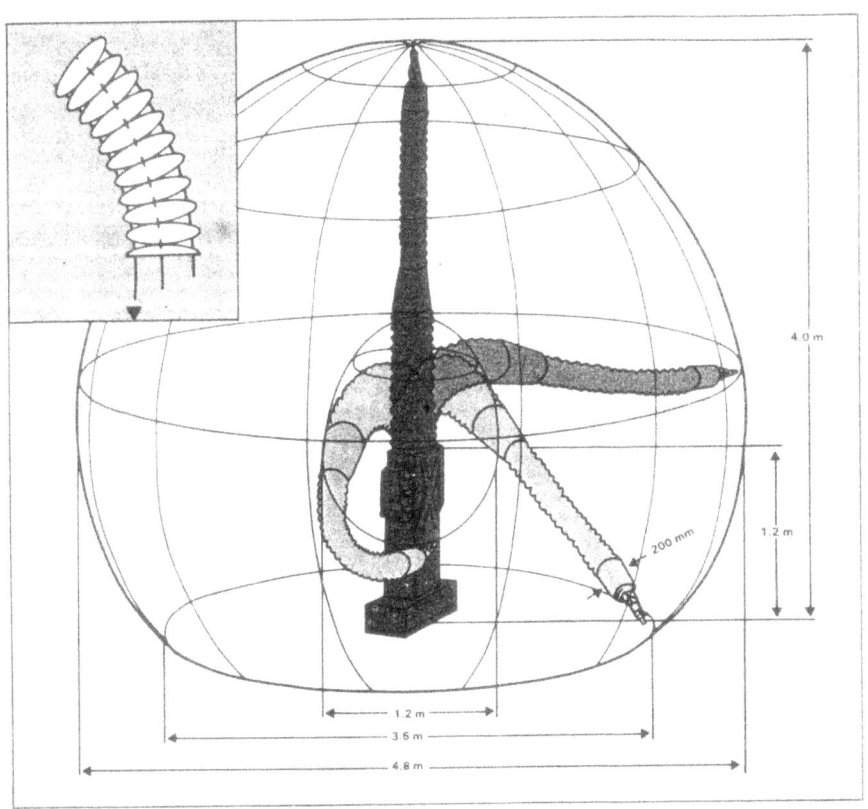

The Adept Three Robot

The Adept Three robot is a four-axis SCARA-type robot with a reach of 42 in. (1067 mm) for its direct-drive unit and a payload of 30 lb. (13.5 kg.).

Below is the picture of an Adept Three robot:-

The SRS-MIA Robot

This is a small industrial robot whose applications include machine loading/unloading, dispensing adhesives and resins, spraying conformal coatings, quality control and automating tensile tester loading.

The picture of an SRS-MIA robot is presented below:-

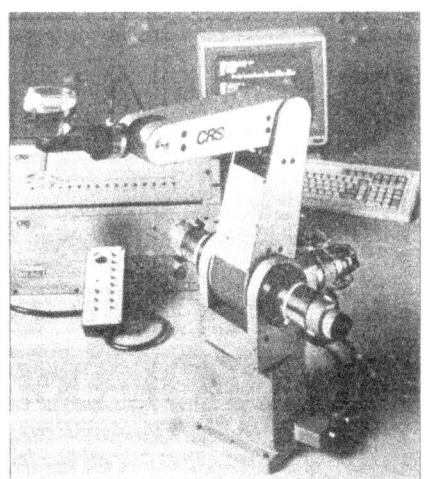

A Changeover Robot

This robot can pick up, use and set down paint spray guns, glue guns, pneumatic or electric power tools and welding tongs. Electric, compressed air or hydraulic power links are produced automatically and four standard and a number of special grinding and deburring tools are available.

The picture of this robot is presented below:-

An Industrial Robot
This is an electric-actuated, instrumented gripper for inserting electronic components or light mechanical assembly. It is equipped with force/torque sensors (for assembly applications), a line of automatic tool-changers and remote center compliance devices. One of the tool-changers is designed for Class 100 clean room use and has a payload of 50 lb. (22.5 kg.) while another has a payload of 150 lb. (67.5 kg.) and weighs about 6 lb. (2.7 kg.).

The picture of this robot is presented below:-

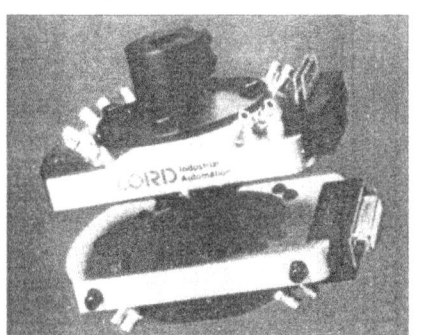

The SSR-H Series Robot

The SSR-H series is endowed with a repeatability of 0.001 in. with an acceleration velocity of up to 79 in. per second and weight-lifting capacity of up to 4.4 lb. and is designed for precision assembly. The robot has a 16-input/16-output interface module which allows it to be controlled by external sensors. To permit interaction with an external computer RS-232C and GP-IB communications interface options are available. The SSR-H series robot is designed for the light machinery and electronics industries and can be used in palletizing, pick and place, screw-driving and accurate assembly, as well as for surface mounting of electronic parts or substrates and insertion of small components on printed circuit boards.

The picture of a SSR-H series robot is presented below:-

The EPR-400 Robot
An electric arc welding robot using the "lead-through teach" method of programming, the EPR-400 is capable of performing MIG, TIG, SAW, PAW and FCAW welding. The operator of the robot provides step-by-step prompting on the CRT display screen. The design of the robot's manipulator is such that it could be mounted overhead when necessary.

The picture of an EPR-400 robot is presented below:-

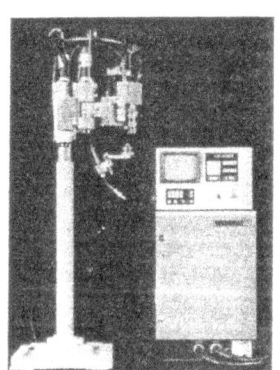

The S-108 Robot
The S-108 robot is designed for arc welding work. Incorporated in the robot are a controller, a positioner, customer-specified welding equipment, a robotic welding gun and a spatter cleaning system. The arc welding software package functions through the robot's controller. The robot's maximum reach is 17 in. and the digitizer has five degrees of freedom. Its perceptor is used in three dimensional computer-aided design, digitization of objects, medicine, contour analysis and field mapping.

Below is the picture of an S-108 robot:-

The System II Robot

The System II robot is designed for material handling applications. With a z axis design it is capable of working in a deep container or on the opposite side of a vertical wall. The robot uses the Cartesian coordinate and can be programmed off line or by the "lead-through teach" method. It can be fitted with a fourth axis to rotate the hand, with several end-effectors available. The x, y and z axes of the robot are DC servo driven microprocessor controlled. Travels of 96 in. in the x axis, 40 in. in the y axis and 66 in. in the z axis are standard. The robot's payload capacity is 250 lb. at 1,500 in. per minute.

The picture of a System II robot is presented below:-